BREWING

A Practical Approach

BREWING

A Practical Approach

BIJAY BAHADUR

Notion Press

Old No. 38, New No. 6
McNichols Road, Chetpet
Chennai - 600 031

First Published by Notion Press 2016
Copyright © Bijay Bahadur 2016
All Rights Reserved.

ISBN 978-1-946204-76-9

Dedication

I dedicate this book to my parents and Purnima, my beloved wife, Tulsi and Parvati, my sisters, and Aditi and Aditya, my wonderful children — all of them have always had the utmost faith in me and given me their unconditional love and support.

Contents

Foreword

I APPLAUD THE publication of your wonderful book on brewing. It is a good reminder to look on the side of the challenges with regard to criticality of the manufacturing of good beer.

I WAS SO impressed that I wanted to thank you personally for your initiatives, knowledge, dedication and professional experience. Of course, much of it is due to your own hard-won experience in the field.

After all the work you put into this publication, it looks like you will finally be recognized. I trust your book will soon be on the bestseller list, and that the success of your first book should keep you going well beyond the second, and third...

P.D. Denzongpa
Managing Director,
Yuksom Breweries Ltd.
Melli, South Sikkim

Preface

THIS BOOK IS for practicing brewers, students pursuing a career in brewing science and consultants who are imparting technical services to the breweries, and have the interest to learn and are determined to expand their knowledge accordingly. It is intended to help brewers and consultants in breweries both large and small. It is not addressed to any specific 'level' of the management.

Practically, it is impossible to describe all aspects of the various types of brewing processes in depth. However, the book can be understood by people working in the alcoholic beverages industry without a degree/diploma in brewing science or technology or even detailed knowledge in these areas. It is not for those who need to be spoon-fed orders and/or instructions, or need a manual to operate the brewery. Everything in this book is entirely in accord with the percepts of brewing experience and personal judgment and dedication.

It took me approximately seven years to complete this manuscript, which found its home at Notion Press Media Pvt. Ltd. I am happy that my ideas, intense study and understanding of the critical aspect of the brewing processes and professional experience have materialized in this book, and hope that each and every one of you that reads it will better understand the philosophy and principles of the brewing processes.

Acknowledgements

FIRST AND FOREMOST, I would like to thank God. In the process of putting together the contents of this book, I realized that God has given me the ability to believe in my passion and to pursue my dreams. I could never have done this without the faith I have in the Almighty.

I would like to express my gratitude to the many people who saw me through this book, to all those who provided support, talked things over, read, wrote, offered comments, allowed me to quote their remarks and assisted in the editing, proofreading and design.

I would like to thank Mr P.D. Denzongpa—Managing Director, M/s Yuksom Breweries Limited, Melli (Sikkim)—for enabling me to publish this book. Without you, this book would never have found its way to so many people.

I want to thank the members of my family, who supported and encouraged me in spite of all the time it took me away from them. It was a long and difficult journey for them.

I would like to thank my mentor, Mr H.A. Devendra Kumar, and my colleagues for helping me in the process of selection and editing. Thanks to Notion Press, my publisher, who encouraged me.

Last but not least, I beg forgiveness of all those who have been with me over the course of many years and whose names I have failed to mention here.

Bijay Bahadur

1

Introduction

THE BREWING INDUSTRY is not large in terms of its contribution to industrial production or total employment. It probably does not employ more than 0.5 percent (approximately) of the working population, nor does the value of its output rise to more than 1.5 percent of the Gross National Product (GNP).

The brewing industry is well organized, and has several international associations. The notable ones are:

1. The European Brewery Convention (EBC), which organizes regular congresses, conducts collaborative research and develop analytical methods (published in *Analytica EBC*)

2. The American Society of Brewing Chemists (ASBC), which is similar to the EBC in terms of functions

3. The Master Brewers' Association of America (MBAA), which is particularly concerned with brewing technology

4. The Institute of Brewing, which exists in Britain and Australasia, and is concerned with collaborative research, developing of analytical methods, technical instruction and publication

A brewer must maintain good control over the quality of his raw materials, brew his beer carefully by using a proven yeast strain, finish it properly and follow good packaging procedure to produce superior quality beer. With all this included, if the brewer's formula can produce beer with a good flavor construction, he will be successful. If he chooses a recipe that is less tasteful, he can expect to attract a smaller but faithful following. However, if he is careless in his brewing and minor flavor variations begin to show up in his product, any success he might enjoy initially will soon be gone. Because they vary in their taste preferences, beer consumers purchase a variety of beers. However, they seem to expect that their chosen beer should present the same flavor at each tasting. If not, they will soon discard it.

Good beer is characterized by a few quality characteristics, which are easy to agree upon. They are:

1. Flavor (combination of taste, aroma and feel)
2. Appearance (color, clarity, foam, and beading)
3. Wholesomeness (absence of hazardous compounds, presence of useful compounds).

Given the choice (which isn't always the case), brewers throughout the world prefer good quality malt—they believe that this will increase their chances of both professional and commercial success. They know that, with the right malt, the job of making good beer will be easier. However, despite (or, perhaps, *because of*) the presence of voluminous literature on the topic, they encounter more difficulties in defining good malt than they do in defining good beer. I plan to address this issue in this book.

Fundamental Brewing Processes

The brewing process/production methods will differ from brewery to brewery, as well as according to brewery equipment and beer types. The main processes will, however, be similar. The description below applies to the production of typical lager beer in a brewery that has a Lauter Tun installed.

Summary

The first stage in the brewing process is the preparation of the wort. After the malt is crushed to grist of suitable fineness, and preliminary treatment of the adjuncts takes place to facilitate extraction, the malt and adjuncts are mixed with brewing water to form a mash. Adjuncts are a supplementary sugar supply, provided either as rice or maize flakes. The mash is heated, following a preset time-temperature program, in order to convert and dissolve substances from the malt and adjuncts in the brewing water (the brewing water is often called 'liquor').

Extraction is accomplished through a combination of simple dissolution and the influence of the enzymes formed during the malting. The substances dissolved in the water are collectively called the 'extract'. The solution formed is called the 'wort'.

When the mashing is completed, the spent raw material, called 'spent grains', is separated from the wort by lautering. The wort is then boiled with hops or hop extracts, releasing bitter substances and oils that are dissolved in the wort. During boiling, the bitter substances are isomerized, which increases their solubility, and a precipitate consisting mainly of proteins is obtained (the 'trub'). After separation of the trub, the wort is cooled to approximately 9–10°C. The cooled wort is transferred to the fermentation vessel for fermentation.

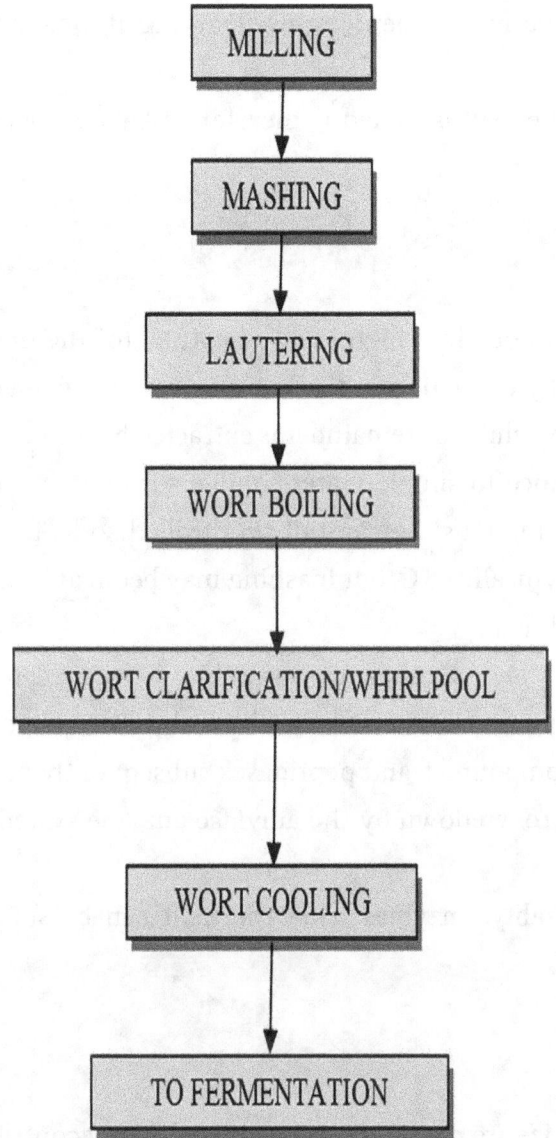

Brewing Process

Milling

In order to obtain a high yield of extracted substances as quickly and efficiently as possible, the malt must first be crushed before it is mixed with hot water. Care should be taken during crushing to make sure that the husks are not damaged, since they are used as a filter bed for separating the spent grains when straining off the wort.

Since the extract yield generally increases in direct proportion to the degree of fineness to which the malt is milled, it is preferable that the malt be crushed to very fine flour. However, this would cause the filter bed to become clogged during most wort straining operations, resulting in an increase in the time required for separation. Moreover, the bed would become less permeable, leading to a hold-up of valuable extract within the spent grains. Sparging, if used, would also be less effective in recovering extract.

The fineness to which the malt is milled is therefore a balance between best extract yield and ability to filter the wort.

Mashing

The purpose of mashing is to obtain a high yield of extract (of the highest possible quality) from the malt grist and adjuncts by extraction in the brewing water. Only a minor part of the extract is obtained through dissolution, while the remainder is extracted by means of the enzymatic breakdown of complex insoluble substance to simple water-soluble substances. Factors such as temperature, pH, and time spent in mashing must be carefully controlled so as to create optimum conditions. Conversion temperature is typically 65°C, but mashing may begin at lower temperatures (45°C) if the malt is relatively under-modified.

Some classes of proteins and starches are insoluble in water. During mashing, the proteins are broken down by an enzyme system. Proteases hydrolyze the proteins into peptides and other less complex nitrogenous compounds, and peptidases subsequently break down the peptides into amino acids. The starch is broken down by the amylase enzyme system into glucose, maltose and dextrins.

In case of poor malt quality, enzymes from the malt can be supplemented with exogenous enzymes.

Adjunct Mash

Adjuncts such as rice and maize are not pre-germinated and do not contribute enzymes. Furthermore, their starch has a higher gelatinization temperature than malt starch. The adjunct mash is, therefore, mixed with water and cooked. The adjunct mash is then mixed with the malt mash, and the malt enzymes break down the adjunct starch. Sugar and glucose syrup can also be used as adjuncts. Since no enzymatic break down is required, these adjuncts are added to the wort kettle (see below).

Lautering/Wort Separation

During mashing, the substance in the malt and adjuncts are broken down and dissolved in the brewing water. In addition to sugar and protein compounds of various complexities, the mash also contains insoluble materials (spent grain). The wort is separated from the spent grains by straining it through a porous filter bed formed by the husks. The residual extract in the filter bed is leached out with sparging water.

The temperature of the wort during straining is about 75–78°C.

Wort Boiling

Following the removal of spent grains, the wort is heated to boiling point in the wort kettle and hops are added.

During wort boiling:

All enzymes are made inactive, to prevent the continued breakdown of proteins during fermentation.

1. The wort is sterilized
2. Bittering of the wort occurs due to isomerization of hop alpha acids
3. Unstable colloidal protein coagulates and precipitates
4. Unwanted flavor components evaporate from the wort
5. The wort is concentrated

The wort is normally boiled vigorously for about one-and-a-half hours with minimum boiling intensity of five to eight percent evaporation per hour.

Wort Clarification/Whirlpool and Cooling

The degree of wort clarity required depends on the type of beer being produced and on brewing practices. The wort should be clear and free of particles ('hot trub') before entering the fermentation vessel.

Particles, especially small ones, carry lipids that, in high concentration, have a marked influence on the formation of ester and higher alcohols, as well as the production of other flavor components from yeast during fermentation.

The equipment most commonly used for wort clarification is the whirlpool, in which the wort and trub particles are introduced in a tangential mode. Secondary forces on the particles cause them to migrate and accumulate in a cone at the center of the bottom of the vessel.

After clarification, the wort is cooled to a temperature of typically about 10°C. Cooling normally takes place in a heat exchanger. The hot water produced is collected and used as brewing water.

Fermentation/Beer Processing Area

The wort is now ready for fermentation and processing. (This process is summarized in the flowchart 'Fermentation'.)

After the wort has been cooled to the fermentation temperature, oxygen is added. The wort is then pumped to the fermentation tanks, where yeast is added.

Oxygen is necessary to support development of the yeast to a state and amount capable of fermenting the wort efficiently. Fermentation is an anaerobic process—the yeast metabolizes the fermentable sugar in the wort, forming alcohol and carbon dioxide.

Heat is generated during fermentation. To maintain the desired fermentation temperature, the fermentation vessel must be cooled.

Once primary fermentation has occurred, the yeast is cropped and pumped into storage tanks. During fermentation, yeast is produced in excess. A part of this yeast is reused for a new batch of wort, with the remainder discharged or treated as a by-product. The production yeast is typically reused several times (sometimes up to ten generations).

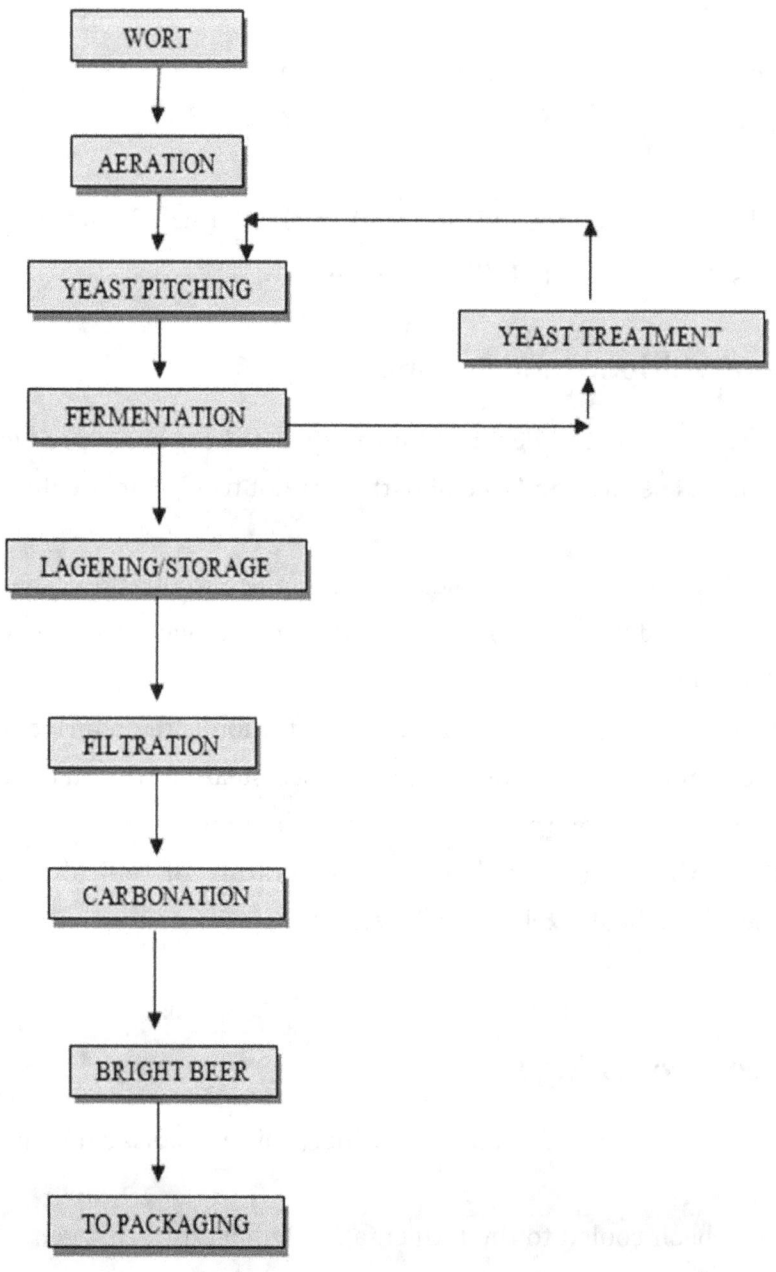

Fermentation / Beer Processing

Yeast Treatment

Yeast treatment involves the following functions:

- Yeast propagation—(i.e.) production of new yeast
- Pitching of yeast to wort
- Storage of production yeast
- Storage of discharge of surplus and spent yeast

Lagering/Storage

The beer is stored for a certain period following primary fermentation.

- Settling of yeast and other precipitates
- Maturation
- Stabilization
- CO_2 saturation

After storage, the beer is ready for filtration.

Pre-filter Cooling

The cooling of beer prior to filtration is important with regard to shelf life, since a forced precipitation of haze particles takes place at this stage (rather than later) in the bottles.

Cooling should be controlled very carefully to ensure a constant beer temperature of approximately -1°C to -1.5°C.

Filtration

The purpose of filtration is to obtain the specified low level of initial haze in the beer and to facilitate prolonged shelf life.

Filtration typically takes place in a *Kieselguhr* (diatomaceous earth) filter. Diatomaceous earth performs the filtration, with the filter itself acting as support for the filter cake.

The small diatoms form a rigid but porous filter cake that sieves out particulate matter as it passes through the filter. To prevent 'blinding' of the filter, and to achieve extended filter runs, Kieselguhr is continuously dosed into the unfiltered beer as 'body feed', thereby constantly building up the depth of the filter cake.

To act as a 'polish filter' after the Kieselguhr, a cartridge filter or sheet filter can be installed.

Additives

Additives such as stabilizing agent, coloring and antioxidant can be added to the beer.

Carbonation

In order to achieve the finished product specification for CO_2, the beer is carbonated before being sent to the bright beer tank. Nitrogen gas may also be used in small quantities to favor foam performance.

Bright Beer Tanks

When the filtration process is completed, the beer is ready for packaging. Prior to packaging, the beer is stored in bright beer tanks.

Cleaning in Place (CIP)

It is important that all process equipment and pipes are kept clean, and disinfected.

Cleaning is done by means of Cleaning in Place (CIP) plants, where cleaning agents are circulated through the equipment or sprinkled over the surface of the tanks. Disinfection takes place through a combination of high temperature, cleaning agents and disinfectants. Caustic and/or acid are normally used as cleaning agents. The cleaning and disinfection of the brewery equipment can involve the use of substantial amounts of energy, water, cleaning agents, and disinfectants.

Several CIP units are usually required in order to cover all the process areas in the brewery.

Packaging

Process Summary

From the bright beer tanks, the beer is pumped to the packaging area, where it is bottled.

During the final operation it is important that:

- The beer is prevented from coming into contact from with oxygen
- No carbon dioxide is lost, as the beer was carbonated to specifications during beer processing
- The beer is not infected (if it is infected, its shelf life will be reduced and the beer will have off-flavors)

The most important functions of a packaging line (for returnable bottles) are shown in the flowchart below.

In packaging lines that use non-returnable bottles and cans, the bottle/cans are flushed only with water before filling.

The bottle washer consumes large quantities of energy, water and caustic. Furthermore, substantial quantities of wastewater are discharged. The use of non-returnable packaging material reduces the consumption of energy, water and caustic, therefore reducing wastewater generation.

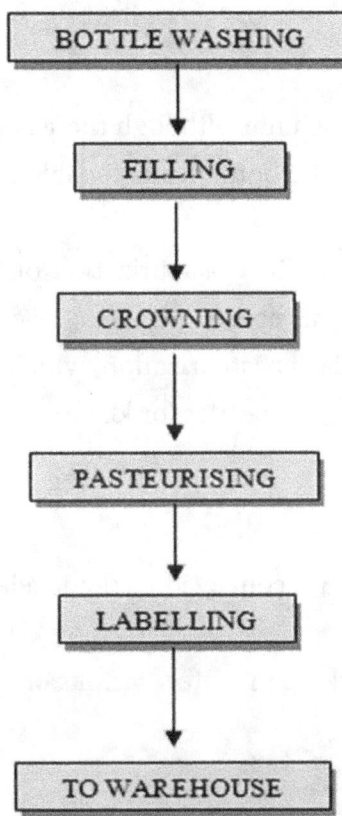

Packaging Returnable Bottles

Bottle Washing

Before being filled with beer, the required bottles are sent to a bottle washer that removes all impurities, inside and outside. Inside the bottles, impurities include residual beer mold, cigarette butts, and so on; externally, impurities may include labels, aluminum foil, and dust particles.

Bottle washing is likely to consist of soaking, rinsing, sterilization and re-rinsing.

Filling

The bottles are transported on conveyor belts from the bottle washer to the filling machine. They are filled under pressure, according to the quantity of dissolved carbon dioxide in the beer.

In addition to filling bottles, the most important function of a filling machine (which may take various forms) is to prevent oxygen from coming into contact with the beer.

Crowning

After filling, the bottles are conveyed to the crowner, which fits them with crown corks. The sealed bottles are then conveyed to the tunnel pasteurizer.

Pasteurizing

Beer is pasteurized to ensure a long shelf life, although there is increasing use of sterile filtration as an alternative. It is important that all microorganisms capable of growing in the beer are destroyed. Pasteurization guarantees biological stability.

Two different methods are used for the pasteurization of the beer: (a) tunnel pasteurization, during which the beer is pasteurized in bottles (or cans)—(i.e.) both the beer and the bottles are pasteurized as a closed unit; and (b) flash pasteurization, which employs a heat exchanger in which the beer is pasteurized before it is put into bottles (or kegs).

Labeling

Following pasteurization, the bottles are conveyed to the labeler, where labels and, in some cases, foil are applied.

The bottles are now ready for packing in crates, cartons or other forms of transport packaging.

Ancillary Operations

Ancillary operations may include:

- Warehouse
- Laboratory
- Boiler plant
- Refrigeration plant
- Water treatment plant
- CO_2 recovery plant
- Compressed air plant
- Electrical supply
- Wastewater treatment plant
- Workshops
- Auxiliary materials storage
- Waste storage
- Residuals handling

Warehouse

Packed beer is stored in the warehouse. It is important to store beer under cover, as sunlight destroys its quality.

In some breweries, there is a centralized warehouse where chemicals and other supplies are kept.

QC Laboratory

Brewing laboratories vary considerably from brewery to brewery. Well-equipped laboratories provide quality control checks on all aspects of raw materials control and use, brewing, fermentation, storage, filtration and packaging. In addition, they can analyze consumer product samples, process water, cleaning water, effectiveness of cleaning procedures, wastewater, and so on, on a routine basis.

Boiler Plant

Heat is supplied by a boiler plant that is usually located on-site. It can take the form of steam or water at high temperatures.

Boilers are normally fired by oil, natural gas or coal, or supplemented with biogas from the anaerobic wastewater treatment plant.

Refrigeration Plant

Process cooling is supplied by a central plant using reciprocating or screw compressors. It can be distributed via the cooling media (for instance, ammonia) directly or via secondary cooling media such as propylene glycol.

Water Treatment Plant

Water can be supplied from wells or surface intake. If the water is supplied from the brewery's own intake, it should be treated in conventional water treatment plants. The water quality must, as a minimum, meet regulatory requirements for drinking water.

The various purposes for which water is consumed in the brewery have special water quality requirements—typically that it be soft or chlorine free. Softening plants are regenerated using either salt or acids. Activated-carbon filters are normally used to remove free chlorine.

Water consumption in breweries varies significantly. Many breweries have installed water reservoirs and use booster stations for local water supply.

CO$_2$ Recovery Plants

The CO$_2$ generated during the fermentation process can be collected, cleaned, and then reused in the brewery. Carbon dioxide is necessary for carbonation and/or counter pressure in tanks and bottles to prevent beer oxidation. Some breweries recover more CO_2 than is needed for production, in which case the surplus can be sold.

Some breweries use nitrogen instead of CO_2 for counter pressure in tanks and bottles.

Compressed Air Plant

Compressed air is mainly used for instruments, actuators, pressurizing of tanks and, possibly, the transportation of spent grain.

Electricity Supply

In India, breweries use electricity from the state's Electricity Board. If the electricity supply is not stable, emergency generators may be used as back-up power supply.

Effluent Treatment Plant (ETP)

The existence of an effluent treatment plant will depend on local discharge requirements and the cost of effluent treatment.

Relevant wastewater treatment technologies involve neutralization and anaerobic or aerobic processes.

Maintenance Workshops

There may be a centralized workshop where spare parts are tooled, equipment and vehicles maintained, and so on; workshops may also be located near various operations.

Auxiliary Materials Storage

Storage for auxiliary materials such as Kieselguhr, glue, labels, and so on, should ideally be located near where the materials are used.

Waste Storage

Solid waste is normally collected in a solid waste area equipped with compartments or containers for each type of waste.

2

Barley Malt

What is malt?

Malt is germinated barley grain that has been dried in a process known as 'malting'. Barley malt grains are partially germinated seeds that have been heated and dried. Contrary to expectation, they are not necessarily dead. Many can re-germinate, given the opportunity (water, air, time), though they are seldom capable of growing into a healthy plant. Malting grains develops the enzymes required to modify the grain's starches into sugars, including monosaccharides such as glucose or fructose, and disaccharides, such as sucrose or maltose. It also develops other enzymes, such as proteases, which break down the proteins in the grain into forms that can be used by yeast.

Malt comes in different styles, depending on how high the kiln temperature is, and the length of time the barley malt stays in the kiln. Generally, malts fall into two broad types—*standard* and *specialty*.

Standard malts, which include those used for the bulk of the grist of both lagers and ales, provide extract, flavor, color and nutrients for yeast.

Specialty malts are used primarily to supply color and flavor, while sacrificing extract yield. The higher the temperature in the kiln, the darker the color and stronger the flavor of the beer.

Pale malt or lager malt is the standard malt used to make most beers. Barley is baked in the kiln over a period about two days, during which time the temperature is slowly raised. This malt is used in light-colored lager beer.

Amber malt—to produce this malt, the barley grain is heated to a higher temperature than pale malt, in order to give the beer a coppery hue.

Crystal malt is used to add a fuller, sweeter flavor to the beer. Green malt is steeped up to forty to forty-five percent water content. Steep malt transfer to roasting drums for up to three hours at temperatures of 60–75°C is maintained under limited airflow. During the drying of the malt, the temperature is raised to 100–120°C to promote Maillard Reactions.

Chocolate malt—kilned malt is transferred to a roasting drum, and the temperature is raised to at least 160°C, and then to over 200°C. A very efficient formation of melanoidins occurs. In the last stage, the temperature is raised to 230°C and maintained until the desired color is reached. This produces a complex mix of roasted flavors as well as a dark color.

Black malt is chocolate that has been taken to near burning point. It has a powerful, bitter taste and is used in small amounts in stouts and porters.

The most important aspect of malt is that it is substantially inhomogeneous. There are considerable differences between individual corns. This has a major impact on processing, and on prediction of the performance of a single batch.

2-Row Barley and 6-Row Barley

Barley varieties are divided into 2-row barley (*Hordeum distichon*) and 6-row barley (*Hordeum vulgare*).

Family	: *Gramineae*
Sub-family	: *Festucoideae*
Tribe	: *Triticeae*
Genus	: *Hordeum*

Brewer's Malt from two Species of Barley:

1. *Hordeum distichon* (2-row barley)
2. *Hordeum vulgare* (6-row barley)

Features: 2-Row Barley

- Traditionally European; dominates the malting barley cultivation in Europe
- Plumper grains; the kernels are larger and plumper due to less crowed ears
- Higher 1,000 corn weight
- Higher extract; higher proportion of starch and less husk
- Lower protein (N × 6.25); less protein content (usually < 1.8% N)

Features: 6-Row Barley

- Traditionally American
- Thinner grains
- Lower 1,000 corn weight
- High losses on screening
- Higher enzyme potential
- Higher protein content (up to 2.5% N)

Benefits: 6-Row Barley

- Used when high enzyme activities in the malt are required, especially when adjuncts such as corn and rice are used
- High protein barley is best suited for animal feed

Why malt?

Barley is difficult to process into wort. It does not mill easily, lacks appropriate levels of enzymes for starch and protein conversion, and the starch is not easily accessible. In addition, it lacks the *malty* flavor. Germination corrects most of these deficiencies, and kilning takes care of flavor. Some beers are produced from barley alone, using exogenous enzymes. The attraction of such beers lies in their favorably low cost, not usually in their quality or ease of production.

How does malt influence the cost of beer production?

Important considerations in selecting malts for use in brewing (from the brewer's perspective) are:

1. Purchase cost of the raw material (barley)—this is driven by yield (which, in turn, is closely linked to disease resistance, need for pesticides, need for fertilizer and so on), climatic and agronomic considerations, and marketplace conditions
2. Cost of malting—attributes that impact malting cost include dormancy, germinative capacity and vigor, need for growth modifiers such as gibberellic acid or bromate and so on
3. Losses in handling—malt damage impacts cost, and results from an interaction between malt quality and handling; the key control parameter is grain moisture content
4. Brewhouse yield—the amount of extract that can be derived from the malt in the brewery

Chemical Composition of Barley Malt

Barley is the world's most nutritional crop and is recommended for beer brewing. This is because barley contains many elements that are rich and healthy and full of energy for brewing yeast. The composition of barley, including the percentage of minerals, gives a general idea about its uses in the brewing industry.

Barley contains:

- Starch   : 62–66%
- Sugars   : 1–2%
- β-glucan   : 3–5%
- Other hemicelluloses (pentosans) : 4–7%
- Cellulose   : 4–5%
- Lipids   : 2–3%
- Protein   : 8–13%
- Amino acids and peptides   : 0.5%
- Nucleic acids   : 0.2–0.3%
- Minerals   : 2%
- Other substances   : 5–6%

What positive attributes does malt contribute to beer?

In addition to providing nutrients for yeast growth, malt contributes a variety of positive characteristics to beer, which are summarized in Table 2.1.

Table 2.1: Positive characters contributed by malt to beer

Character	Compounds responsible
Malty flavor character	*O*- and *N*- heterocycles
Sweet corn flavor character	Dimethyl sulfide (*syn* methyl sulfide)
Flavors due to esters and higher alcohols	Malt provides the precursors for these materials (amino acids and sugars)
Color	Melanoidins and polyphenols
Foam	Polypeptides, some of which are glycosylated

What negative attributes can malt contribute to beer?

Negative characteristics contributed to beer by malt are summarized in Table 2.2. In addition to the effects on beer, malt components can also influence the brewing process. For example, excessive levels of beta-glucan can give rise to poor wort filtration and beer filtration performance. Malt components can bring about premature flocculation of bottom fermenting yeast during fermentation, or changes in the skimming behavior of ale yeast strains. Also, malt may be implicated in fobbing during fermentation.

Table 2.2: Negative characteristics contributed by malt to beer

Character	Compounds responsible
Off-flavors (grainy, astringent, stale, etc.)	Various small molecules (including *O*- and *N*-heterocycles, carbonyls, etc.) and enzymes (including lipoxygenases, lipases, etc.)
Safety concerns (toxin, mutagen, carcinogen or teratogen)	*N*-Nitrosodimethylamine, chloropropanols, aflatoxins, mycotoxins
Appearance problems (haze)	Proteins, polyphenols
Foam problems (gushing)	Polypeptide derived from fungal contaminants on grain

3

Adjuncts

Introduction

A brewing adjunct is defined as unmalted grains (such as corn, rice, rye, oats, barley and wheat) or grain products used in brewing beer, which supplement the main mash ingredient (such as malted barley). Adjuncts are used mainly because they provide extract at a lower cost (a cheaper form of carbohydrate) than that available from malted barley, but are sometimes to create additional features, such as better foam (head) retention, flavors or nutritional value or additives.

A wide range of materials fall within this definition and, in this chapter, attention will be directed to three areas: (a) solid unmalted raw materials usually processed within the brewhouse; (b) liquid adjuncts usually added to the kettle and some specialty products used for priming; (c) malted cereals other than barley, such as wheat and sorghum (Table 3.1). Adjuncts are usually considered as non-malt sources of fermentable sugars. They typically contribute no enzyme activity and little or no soluble nitrogen, and are less expensive than malt. It is also sometimes stated that adjuncts do not contribute flavor to the finished product; it will be discussed later that, however, in many beers, this is not really the case.

Table 3.1: Brewing Adjuncts and Their Preparation Processes, in Increasing Complexity

Basic raw cereal	: Barley, wheat
Raw grits	: Corn (maize), rice, sorghum
Flaked	: Corn, rice, barley, oats
Torrified/micronized	: Corn, barley, wheat
Flour/starch	: Corn, wheat, rice, potato, cassava, soya, sorghum
Syrup	: Corn, wheat, barley, potato, sucrose
Malted cereals other than barley	: Wheat, oats, rye, sorghum

Purpose of using Adjuncts:

The use of adjuncts in brewing recipes has become common since the early 1900s for some practical reasons:

1. To reduce the size of the protein fragments in all-malt beers so that they would not become turbid
2. Beer drinkers' preferences leaned towards the lighter palate of beers with lower malt usage
3. Cost reduction (economy)—reduces malt cost (some adjunct provides cheap extract), increases capacity and energy savings, reduces demand of mashing equipment and aids high-gravity brewing
4. Change beer characteristics such as color, foam or flavor

Types of Brewing Adjunct:

- Basic Raw Cereal : Barley, wheat
- Raw Grits : Corn (maize), rice, sorghum
- Flaked : Corn, rice, barley, oats
- Torrefied/Micronized : Corn, barley, wheat
- Flour/Starch : Corn, wheat, rice, potato, cassava, soya, sorghum
- Liquid : Glucose syrup, sucrose syrup, invert sugars, malt extract, caramel, priming sugar
- Malted Cereals Other than Barley : Wheat, oats, rye, sorghum

Adjuncts Provide:

1. Starch—provides cheap extract
2. Protein—some adjunct dilutes the protein level, which has an effect on beer stability (for example, wheat glycoprotein stabilizes foam, due to the wheat gluten's effect on filtration problems)
3. Beta-glucans, pentosans—highest in barley and lowest in sorghum; β-glucan from sorghum is highly soluble and, hence, causes difficulties in filtration
4. Lipids—some adjunct contributes more fat, which results in 'stale' flavors, and causes foam to collapse
5. Inorganic materials such as Zn, Cu, Ca, Mg, Fe, P and S—may have meaning as growth factor of some yeast strains, but balance is important
6. Color—some adjuncts can be used to improve beer color

A great deal of effort has been expended to improve the performance of various adjuncts and to examine their contribution to the characteristics of the finished beer. In general, corn tends to give a fuller flavor to beers than wheat, which imparts dryness. Barley will give a stronger, harsher flavor. Both wheat and barley adjuncts can considerably improve head retention (foam). Rice will also give a very characteristic flavor to beer.

The overall brewing value of an adjunct may be expressed by the following equation:

Brewing value = Extract + Contribution to beer quality - Brewing costs

The major benefit is extract.

In India, current use of non-malt adjuncts averages about thirty to thirty-five percent of the total brewing materials employed, excluding hops. The most commonly used adjunct materials are corn (maize), rice, barley, and sugars and syrups. The following materials are used as unmalted brewery adjuncts: yellow corn grits, refined corn starch, rice, sorghum, barley, wheat, wheat starch, cane and beet sugar (sucrose), rye, oats, potatoes, tapioca (cassava), and triticale. In addition, processed adjuncts include corn, wheat and barley syrups, torrified cereals, cereal flakes, and micronized cereals.

Corn Grits

Corn grits, produced by dry-milling yellow corn, are the most-widely used adjunct. The milling process removes the hull and outer layers of the endosperm along with the oil-rich germ, leaving behind almost pure endosperm fragments. These fragments are further milled and classified according to the brewer's specifications. Corn grits produce a slightly lower extract than other unprocessed adjuncts and contain higher levels of protein and fat. The gelatinization temperature range for corn grits (62–74°C) is slightly lower than that of rice grits (64–78°C).

Pilot brewhouse studies revealed high extract yields for the flaked and micronized flaked grits, and they showed slightly higher fermentabilities.

After reviewing the composition of worts and beers prepared with a variety of adjuncts, it has been found that a carbohydrate profile similar to an all-malt wort can be achieved with either twenty percent rice or twenty percent corn grits, although levels of sucrose and fructose decline as the adjunct level increases. Corn grits at thirty percent level produce a volatile aroma compound similar to that of an all-malt beer. Wort protein, peptides, free amino acids, and nucleic acid derivatives decline in proportion to the adjunct level. The amino acid profile of wort is not affected by a particular adjunct but by its level in the mash. High adjunct ratios lead to higher levels of diacetyl and related compounds at the end of fermentation. However, with the appropriate post-fermentation processing, these levels return to normal after aging.

Rice

Rice is currently the second most-widely used adjunct material. On an extract basis, it is approximately twenty-five percent more expensive than corn grits. Brewing grade rice (brewer's rice) is a by-product of the edible rice-milling industry. Hulls are removed from paddy rice, and the resulting brown rice is then dry milled to remove the bran, aleurone layers and germ. The objective of rice milling is to remove these fractions completely, with a minimal amount of damage to the starchy endosperm, resulting in whole kernels for domestic consumption. However, up to thirty percent of the kernels are fractured in the milling process. The broken pieces (broken) are considered esthetically undesirable for domestic use and are sold to brewers at a price that is considerably less than that of whole kernel or mill-run rice price. Rice is preferred by some brewers because of its lower oil content compared to corn grits. It has a very neutral aroma and flavor and, when converted properly in the brewhouse, yields a light, clean-tasting beer.

The quality of brewer's rice can be judged by several factors, including cleanliness, gelatinization temperature, mash viscosity, mash aroma, moisture, oil, and ash and protein content. Rice should be free of seeds and extraneous matter. Insect or mold damage should not be tolerated, as these indicate improper storage or handling conditions. It has been reported that rancidity in rice oil can be a problem, but with modern storage techniques this is a negligible factor. Laboratory mashes of rice samples should be conducted regularly and they should gelatinize and liquefy in a standard manner and should be clean and free from undesirable odors and tastes.

Not all varieties of rice are acceptable brewing varieties. Rice has a relatively high gelatinization temperature and is extremely viscous prior to liquefaction in the cereal cooker. Many rice varieties will not liquefy properly and are impossible to pump from the cooker to the mash mixer whereas other varieties liquefy well in the cooker during a fifteen-minute boil. Both the amylose and liquid content of rice varies with the variety and the cultivation conditions—thus, selection of suitable grades is important. Rice liquefies more easily the finer the particle grind, and particles less than 2 mm are considered adequate. Handling of rice is relatively easy, as the broken contain little dust and flow easily through standard hopper bottoms and conveyoring equipment. Rice is milled in fixed roller mills. There is no difficulty in making the rice mash slurry at 64–78°C, although it is a common practice to mash and hold at 36–42°C as a protein rest. As with all cereal cooker operations, whatever the starch source, five to ten percent of the malt grist is added to the cooker because the malt enzymes (amylases and proteinases) are essential for the partial liquefaction necessary to render the cooker mash fluid enough for pumping. Atmospheric boiling is required for gelatinization. Some brewers pressure-cook it at 112°C.

If properly converted, rice adjunct usage does not create run-off problems. As previously discussed, the extract is slightly lower in soluble nitrogen than corn grits.

Barley

Unmalted barley is an obvious adjunct for use in brewing. However, the raw grain is abrasive and difficult to mill, scattering to yield too high a percentage of fine material which gives problems during lautering. These difficulties disappear if the grain is conditioned to eighteen to twenty percent moisture prior to milling although this process has not been widely employed in brewing.

In the past, barley was normally partially gelatinized before use. The barley was gelatinized either by pressure-cooking it mildly or by steaming it at atmospheric pressure followed by passing the hot grits through rollers held at approximately 85˚C. Finally, the moisture content of the flakes is reduced to eight to ten percent. This process of pre-gelatinization can also be applied to corn. Pre-gelatinization of barley affects the ease of extraction of β-glucan during mashing and, hence, the β-glucan content of the wort. Prolonged steaming prior to rolling the barley produces a product that produces higher-viscosity, sweet worts. This can be controlled by measuring the viscosity of a cold water extract of the flaked barley—which is a good indication of the extent of the steaming process. Barley starch is more readily hydrolyzed than corn or rice starch. Barley may be de-husked before use to increase extract yields, but this may lead to run-off difficulties because the husk provides material for filter bed formation. In the same way, fine grinding improves extraction efficiency but also leads to slow run-off. If significant proportions of barley are used in the mash, malt with sufficient enzyme activity is required. Use of barley leads to a reduction in wort nitrogen content and decreased wort and beer color. No difficulties have been reported in fermentation. Foam stability is usually improved because of lower levels of proteolysis. However, a major difficulty associated with brewing with high levels of unmalted barley can be the increase in wort viscosity and run-off times caused by the incomplete degradation of β-glucans. Mashing at 65˚C quickly destroys the malt β-glucanase activity. Suggestions to alleviate these problems have included pre-treatment of the barley with β-glucanase and the use of a temperature-stable β-glucanase in the mash.

Raw (feed) barley can also be employed as an adjunct, and as high as fifty percent barley in the grist has been employed in some breweries in Australia. Use of raw barley requires significant modification to the brewing process. For reasons already discussed, conventional roller milling cannot be employed; consequently, hammer mills are necessary. This high level of malt replacement usually results in insufficient malt enzymes for the necessary hydrolysis of the starch, protein, and β-glucans. Consequently, a malt-replacement enzyme system is employed to compensate for the reduced level of malt enzymes. A number of such enzyme systems have been developed and are usually a mixture of β-amylase, protease, and β-glucanase, which are obtained from microbial sources such as *Bacillus subtilis.*

In barley brewing, it is possible to approximate the starch hydrolysis profile and the degree of fermentability of hundred percent malt worts. This is possible by substituting malt with barley at levels of fifty percent (extract basis) and by controlling the main mash schedule (enzyme concentration, time, and temperature). Barley worts have been found to contain less fructose, sucrose, glucose, and maltotriose but more maltose than malt worts. No anomalies or difficulties in fermentation and aging have been noted. Most breweries can employ their normal fermentation and aging technology for barley brewing. In general, no significant difference in organoleptic properties between barley beers and hundred percent malt beers have been observed. A harshness of barley beers can be avoided by lowering the pH of the wort to 4.9 prior to boiling.

It would appear that, with the aid of microbial enzymes, today's brewer could increase the level of unmalted raw barley. To do so would be to their economic advantage and, at the same time, help obtain the desired beer quality.

Sorghum

Sorghum is the fifth most-widely grown cereal crop in the world (wheat, corn, rice, and barley are produced in greater quantities). Africa is a major source of sorghum, as is Central America. Sorghum is the traditional raw material used in Africa for the production of local top-fermenting beers (that are known by various names). These beers are produced without hops—they are slightly sour in taste, and are consumed unfiltered, mainly in rural regions.

The American brewing industry employed sorghum as an adjunct in 1943 when brewing materials were scarce. Unfortunately, sorghum was cracked and only partially dehulled and degerminated; consequently, brewers obtained poor yields and bitter-tasting beers and faced a number of other quality related problems. Modern milling techniques and better purification methods have changed the situation. Today, sorghum brewer's grits are considered by many to be of comparable quality to the best corn and rice grits.

In Africa and Mexico, brewers are using an appreciable and a continuously increasing percentage of brewer's grits and, in most cases, producing beer of acceptable quality.

In the brewing process, the dried sorghum is screened to remove extraneous material. The whole grain is then fed into a series of de-hullers that produce two product streams. In one of these streams, the husks and embryonic material constituting some forty-eight percent of the original sorghum are removed and this fraction, together with the initial screenings, is sold as a by-product. The second stream, consisting of peeled sorghum together with a small amount of husk material is then passed through an aspirator in which the husk is removed. The purified, pearled sorghum, now representing forty-seven percent of the original cereal processed, is milled to give twelve percent of the original material as flour and thirty-five percent as sorghum grits. Both of these components are used as brewing adjuncts contributing up to forty-five percent of the total wort extract.

The chemical composition of sorghum grain is very similar to corn. Both grains contain starch consisting of seventy-five percent amylopectin and twenty-five percent amylose. Starch granules are similar in range, shape, and size. On an average, sorghum starch granules are slightly larger—15 mm when compared to 10 mm for corn. Sorghum starch has a higher gelatinization temperature (68–76°C) than corn starch (62–68°C). In the brewhouse, sorghum brewer's grits perform within acceptable limits. No special handling or cooking techniques are required. Five percent malt in the cooking mash is sufficient. Conversion of starch occurs within the mashing time allowed. The beers produced are fully equivalent in chemical analysis, flavor, and stability to beers produced with other adjuncts. Finally, in many areas (such as Africa and Central America), sorghum offers the lowest-cost source of available fermentable sugar.

Refined Corn Starch

Refined corn starch is by far the purest starch available to the brewer. It is a product of the wet-milling industry. It has not found widespread use because its price is higher relative to corn grits and brewer's rice, and is difficult to handle. An obvious drawback for refilled starch usage is handling. The starch powder is extremely fine, with ninety-six percent passing through a 200-mesh screen. It must be contained in well-grounded lines and tanks to prevent explosions resulting from static electricity sparks produced during conveying. The starch bridges easily and is nearly impossible to flow from tanks unless they have special fluidizing bottoms.

Refilled corn starch can be utilized as a total adjunct or can be mixed with rice or corn grits as per the brewer's choice. The gelatinization and liquefaction of the starch proceeds at lower temperatures than rice or grits, but is not sufficiently different to preclude its use as a blend with either. As a total adjunct, care must be taken to prevent it from sticking to the cooker bottom.

Refilled starch can easily be liquefied by the same process utilized for rice. The resultant extract cannot be organoleptically or chemically differentiated from an all-rice extract. Brewhouse yield can be increased by one to two percent by the use of refilled starch in the place of rice. There are no run-off problems. Fermentations tend to attenuate better, while colloidal stability is unaffected. Beer flavor is not affected, except that the beer is considered slightly thinner, because of higher attenuation limits.

Wheat Starch

Refined wheat starch is not presently attractive because of its high price compared with the more readily available adjuncts. Chemically, wheat starch is very similar to refilled corn starch. An advantage is that its gelatinization temperature is similar to malt gelatinization and could be added directly to the malt mash; however, ten percent higher brewhouse yields can be obtained by cooking in a conventional adjunct cooker. Lautering times are reported to run up to ten percent longer than with corn grits.

Wheat starch has the same conveying and handling problems as refined corn starch. Slurrying should take place below 52°C to prevent lumping. The cooker temperature should not exceed 98°C, as the starch foams badly upon boiling.

Wheat starch is somewhat higher in β-glucans and it is suggested that the cooker mash, with ten percent of the malt added, stand at 48°C for thirty minutes prior to the 66°C rise to give the β-glucanase time to break down the β-glucans at its optimal temperature. This procedure results in little or no run-off problems.

The beer is quite comparable to the beer brewed with corn grits in analysis and flavor. Should wheat starch be available at prices competitive to other adjuncts, it would be a perfectly suitable adjunct.

Torrified Cereals

Torrification is a process by which cereal grains are subjected to heat at 260°C and rapidly expanded or popped. This process renders the starch pre-gelatinized and thereby eliminates the cooking step in the brewhouse. It also denatures a major portion of the protein in the kernel such that the wort soluble protein is only ten percent of the total.

Both barley and wheat are potential materials for torrification and for use as torrified adjuncts. The chemical analyses are quite similar for both. The torrified products have about 1.4 percent wort-soluble protein and could allow the use of lower protein malts or higher adjunct levels, while maintaining soluble protein similar to worts produced with lower soluble protein adjuncts. Fat content is slightly higher than for other adjuncts, but again this would be negated in the final wort by using higher adjunct levels.

There are no handlings or dust problems associated with the use of torrified cereals. It is possible to blend the torrified products with malt. They can then be ground simultaneously and mashed-in together. However, higher yields are found by cooking the torrified product separately at 71–77°C, prior to addition to the malt mash.

The use of torrified cereals leads to increased lauter grain bed depth and to slight run-off penalties. Torrified barley seems to be more refractory than torrified wheat in this respect. Particle size and mill setting are critical; large particle size leads to poor yield, and too fine a grind leads to run-off problems. Because of its expanded nature, the torrified cereals absorb more water than other adjuncts and, especially in the case of torrified barley, higher ratios of water to cereal must be used. The flavor of beer produced with torrified adjuncts is reported to be unchanged, and one could easily conclude that if torrified cereals become economically competitive with other adjuncts they would be employed as an alternate adjunct source.

Liquid Adjuncts

The major liquid adjuncts used in brewing are glucose syrups, cane sugar syrups, and invert sugar syrups. Glucose is the commonly used name for dextrose, but the glucose syrups used in brewing are, in fact, solutions of a large range of sugars and will contain, in varying proportions—depending upon the method of manufacture—dextrose, maltose, maltotriose, malto-tetraose and larger dextrins.

Cane sugar syrups contain sucrose derived from sugar cane and sometimes, depending upon the grade, small quantities of invert sugar. Invert syrups, as the name suggests, are solutions of invert sugar—a mixture of glucose and fructose. Invert sugar is produced, in nature and commercially, by the hydrolysis of sucrose, which, together with glucose and fructose, occurs abundantly in nature. Commercially, sucrose is extracted from sugar cane or beet, and glucose syrups are usually manufactured from starch derived from corn or wheat grains.

Glucose syrups have been available since the mid-1950s. They were originally produced by straight acid conversion of starch to a 64–68 DE range. (The degree of starch conversion is usually expressed as 'dextrose equivalent' or DE. This is a measure of the reducing power of the solution, expressed as dextrose in dry solids.) The brewer's main concern is that the apparent extract of the finished beer should not change with the addition of the liquid adjunct, that is, the syrup has to be approximately twenty percent non-fermentable. The high level of glucose became a concern with the use of liquid adjuncts.

Table 3.2: Sugar Spectrum (%) of First and Second Generation Liquid Adjunct (Acid and Acid/Enzyme Converted, Carbon Refined), Third Generation Liquid Adjunct (Enzyme/Enzyme Converted, Ion Exchanged) Compared to All-malt Wort

	Liquid Adjunct (First generation)	Liquid Adjunct (Second generation)	Liquid Adjunct (Third generation)	All-malt Wort
Glucose	65	40	5	8
Maltose	10	28	55	54
Maltotriose	5	12	20	15
Dextrin	20	20	20	23

Conversion of starch with the aid of an acid produces predominantly glucose as the hydrolysis product. When brewer's yeast is exposed to high concentrations of glucose, a phenomenon referred to as the 'glucose effect' may be experienced, this can result in sluggish and 'hung' fermentation.

Malt from Cereals Other than Barley

Although the principal cereal employed as the raw material for malt is barley, a number of other cereals are used, including wheat, oats, rye and sorghum.

Wheat Malt

Wheat malt is used in the production of some special types of beer, such as Berlin Weiss beer, in which it may constitute seventy-five percent of the grist, but only to a limited extent in ordinary beers. The limited use of wheat malt is mainly due to the difficulty experienced in malting the naked grain without damage to the exposed acrospires. As a result, much of the wheat malt made has been under-modified. However, the absence of the husk tends to result in a high extract. Wheat malt gives beer outstandingly good head retention.

Oats and Rye Malt

Malted oats are used to a limited extent and in some stout are blended with barley malt. Malted rye does not seem to be used today, although some fifty years ago it was used in specialty beers. Unmalted rye is sometimes used for vinegar brewing and also for certain distilled beverages (such as Canadian rye whiskey).

Sorghum

The use of unmalted sorghum has already been stated above. A relatively recent development in the use of sorghum in brewing is as malt. For many years in southern Africa, malted sorghum has been used to brew a traditional alcoholic beverage of the region, known as sorghum or opaque beer. Sorghum beer is characterized by its sour taste—it is flavored by lactic acid produced by bacterial fermentation and is not hopped. Sorghum beer is of moderate alcohol content (approximately three percent w/w) and the opaque appearance is due, in part, to incomplete hydrolysis of starch during mashing. The presence of high levels of complex carbohydrates in sorghum beer makes it a nutritious beverage as well as an alcoholic drink.

Malted sorghum differs in many aspects from barley malt, particularly in terms of the properties of its starch and diastatic enzymes (refer to Table 3.3). Sorghum malt starch has a gelatinizing temperature in the range of 64–68°C, some 10°C higher than that of barley malt starch. The total diastatic activity of sorghum malt is less than half that of barley. This is probably because of the low β-amylase activity of sorghum malt. However, the α-amylase activity of sorghum malt is slightly higher than barley malt.

The development of sorghum for use as malt in conventional beer (ales, lagers, and stouts) production is in the process of rapid development. This development has been accelerated by the large foreign debt crisis in developing tropical countries, which has made it increasingly difficult for them to import either barley or barley malt for their existing breweries. For example, the government of Nigeria has prohibited the import of barley malt since 1988. In order to reduce this economic difficulty, considerable research into local raw materials has been conducted (especially sorghum), not only for their use as adjuncts, but also as a complete replacement for barley malt. As a result of these research efforts, as much as thirty percent of the sorghum harvest in Africa is used for malting and brewing.

Table 3.3: Comparison between Sorghum and Barley Malt

	Sorghum Malt	Barley Malt
Starch gelatinization (°C)	64–68	55–59
Diastatic power	19	53
β-Amylase activity (%)	18	100
α-Amylase activity (%)	110	100

There is general agreement that the 'white' sorghum types are more suitable than the 'red' types because of their lower content of polyphenols. During the malting of sorghum, the pattern of endosperm enzyme breakdown is different from that of malting barley. For example, although malting sorghum grains develop significant levels of endo-β-1,3-glucanase and pentosan enzymes, production of the cell wall-degrading endo-β-1,3:1,4-glucanase enzyme is significantly lower than the levels found in malting barley. Also, as previously discussed, sorghum has a lower enzyme complement, especially β-amylase and endo-β-glucanases, and even with the addition of these enzymes exogenously, starch extracts comparable with barley malt are rarely obtained. Whereas mashing of barley malt at 65°C allows both starch gelatinization and enzyme solubilization of starch to occur simultaneously, mashing of sorghum malt at this temperature fails to gelatinize its starch, and despite high b-amylase activity, starch solubilization and extract development are still inadequate. Finally, the use of sorghum for lager and stout beer brewing will alter its taste and consequently, in many parts of Africa, the consumer has become acclimatized to a different type of beer.

Conclusion

The use of unmalted carbohydrates or adjuncts in brewing is widespread, except in those countries that adhere to the German Purity Law. In most countries, there are one or two dominant adjuncts and these are usually the cheapest suitable carbon source. In Africa, corn remains the most popular adjunct; however, the use of sorghum (as an adjunct and malt) is increasing. Although developments in the use of brewing adjuncts have been relatively stable for a number of years, the advent of 'new-generation' syrups (produced principally, but not exclusively, from corn) is currently having a great impact on some parts of the brewing industry. Biotechnological advances such as wet milling, immobilized thermo-tolerant enzyme systems, and ion exchange downstream-processing techniques permit the production of syrup with virtually any carbohydrate profile. At the present time, syrups are available that allow the brewer to introduce them at any level without changing the carbohydrate profile of the wort. The future will see the commercial ability to separate and isolate individual sugars according to their molecular weight and, subsequently, produce blended syrup of any sugar profile.

4

Hops

HOPS (HUMULUS SPECIES) are the female flowers of a hop species. Hop products are often used as a flavoring and stability agent in beer, to which they impart a bitter, tangy flavor. Hops are sometimes added to some varieties of kvass. They are also used for flavor in some tisanes. Hop products are rather costly, with prices increasing year after year. Controlling its cost is the first job that the brew master should focus on!

Hop Constituents

Hops contain hundred of components, but most of the brewing value is found in the resins and hop oils.

Hop Resins

Hop resins are subdivided into hard and soft, based on their solubility. Hard resins are of little significance, as they contribute nothing to the brewing value, while soft resins contribute to the flavoring and preservative properties of beer. Alpha and beta acids are two compounds present in the soft resins and are responsible for the bitter taste. Alpha acids are responsible for about ninety percent of the bitterness in beer. Magnesium, carbonate, and chloride ions can also accentuate hop bitterness.

Alpha Acids

Alpha acids are the precursors of beer bitterness since they are converted into iso-alpha acids in the brew kettle. The three major components of alpha acids are humulone, cohumulone, and adhumulone.

Isomerization: When hops are added to the boiling wort in the kettle, their alpha acids go through a chemical change known as isomerization.

Beta Acids

Hops also contain a second group of acids known as the beta acids. These (lupulone, colupulone, and adlupulone) are only marginally bitter.

Hop Oils

Hops that have high alpha-acid content are preferred for their bittering and flavoring properties. Hops are also selected for the character of their oils. The oils are largely responsible for the characteristic aroma of hops and, either directly or indirectly, for the overall perception of hop flavors. Hops selected for the character of their oil content are often referred to as aroma or 'noble' type hops. Oils also tend to make a beer's bitterness a little more pronounced, and enhance the body or mouth-feel of the beer.

Hop Varieties

Although there is only one hop species (*Humulus lupulus*) that is used for brewing beer, there are a number of varieties (technically known as 'cultivars') in that species, each with its own spectrum of characteristics. Varied hops are chosen for the bitterness, flavor or bouquet that they will lend to the beer. Hop varieties can be roughly divided into two classes—bittering hops and aroma hops—although there are hops that can be considered dual-purpose.

Bittering Hops

As their name implies, bittering hop varieties are those that impart a bitter flavor to beer and have high alpha-acid levels. Bittering hops usually have a high alpha-acid content.

Aroma Hops

Aroma hops, with low to medium alpha levels, mainly impart characteristic hop aromas to beer. In Europe, emphasis is on growing aroma hops, with a smaller but growing leaning towards bittering hop varieties.

Hop Products

Besides whole hops (hop cones), other hop products used in brewing beer fall into three general categories:

1. Non-isomerized hop products

The alpha acids are unchanged during processing non-isomerized products. As a result, the products can only be added to the kettle only during wort boiling.

2. Isomerized hop products

With isomerized hop products, the alpha acids have been isomerized to iso-alpha acids during processing. As a result, they can be added to the kettle not only during wort boiling, but also—unlike alpha-based products—during the maturation phase.

3. Hop oil products

Pure hop oils usually have a very pale green/yellow appearance. They have been on the market for many years, and give beer hop aroma without imparting any bitterness.

Hop Stability

Hop components (such as alpha acids and essential oils) undergo changes as soon as they are harvested. The rate of loss is dependent in part on storage conditions, and on the variety and condition of the hops when baled. For example, alpha-acid stability is expressed as the percentage of alpha acids remaining in baled leaf hops after six months of storage at ambient temperature.

Although hops deteriorate with storage, the aging process can be slowed with refrigeration, anaerobic packaging, and minimizing contact with light. Of these three, refrigeration is the most important, followed by anaerobic packaging. For optimum preservation, hops should be stored at temperatures between $0-2°C$.

There are a number of things the brewer needs to consider when selecting the hops. He/she needs to know the alpha-acid 'rating' for the specific hops considered for the beer type and style. This rating describes how much of the hop, weight-wise, is made up of alpha acids. Hops with a higher alpha-acid content will add more bitterness than a low alpha hop (when using the same amount).

Each variety of hops contributes a bitterness that balances the sweetness of the malt in the beer recipe, so that brewer needs to choose wisely. In addition, each variety of hops will add a specific aroma or flavor to the beer.

Three different types of product are available for the hop addition;

1. Whole hop (i.e.) most natural, unprocessed form of leaf hops
2. Hop powder products (powder and pelleted hop)
3. Hop extracts

Points to consider with regard to hop addition:

1. In how many parts should the additions be made?
2. When should the partial additions be made?
3. Which varieties first, and which ones at the end?
4. Which aroma hops are to be added last?

Hop products requirement depending on:

- Wort volume

- Bitterness specification (target BU in ppm)

- Hop product(s) to be used

- Perceived bitterness of iso-alpha in product

- Alpha or iso-alpha percentage in product(s)

- Time of addition of each product (at the beginning of boiling, or later addition)

- Estimated utilization in final beer of each addition

Hops in the Brewing Process

Hops give beer a characteristic aroma and flavor. The hopping of a beer has always been part of a formula that is the brewer's secret. The influence hopping has on the taste is also dependent on the *Beer Matrix*—meaning that the same hop formula for a dark and a pale beer will be experienced differently by our senses.

For this reason, it is difficult to know the influence of the hop components on the beer. No instrument can reproduce our sensory perceptions and it is general knowledge that subjective feelings are difficult to describe. Some substances that the hops bring into the beer have sensory thresholds that are below the parts per billion or ppb range ($10-9$), and which are difficult to identify even with the most modern apparatus.

The hop components react just like the beer:

- Sensitive to oxygen

- Sensitive to temperature and change of temperature

- Sensitive to light

- Sensitive to time (aging)

The brewer has a wide range of choices available and the respective use dependent on the brewer's goal:

- To impart a 'basic bitterness' in the beer with selected bitter hop varieties

- To achieve a balanced aroma as a 'flavor profile' in the beer with classic aroma hops

- To impart physiologically valuable tannins to the beer for 'digestibility' with special aroma hops

Traditionally, hopping takes place in the wort kettle. The time for the dosage varies from the first wort hopping, via the additions directly before casting, up to the direct aroma hopping in the whirlpool. After being dosed, the hop components pass through the following stages:

1. Mechanical distribution
2. Emulgation—(i.e.) dissolving the components
3. Transforming components by heat
4. Evaporating volatile (partly undesired) hop oils
5. Partially combining with proteins and precipitation

At the same time, the following processes occur during the wort boil:

- Bitter substances—alpha isomerizes to iso-alpha-acid, dissolving other components of the resin fraction, such as the humulinic acids
- Aroma substances—dissolving, evaporation, oxidation
- Polyphenols—dissolving, polymerizing, precipitation

The iso-alpha acids are sensitive to oxygen and form an unspecific amount of oxidative products while the wort is boiling. Parameters that influence the isomerization are:

- Boiling time
- Temperature/pressure
- pH
- Original gravity
- Wort turbidity
- Level of bitter substances
- Design of wort kettle
- Hop product

Hop Utilization

Hop utilization is the percentage of alpha acids that is isomerized and remains in the finished beer. The utilization of the bitter substances rarely exceeds forty percent in commercial breweries and is often as low as twenty-five percent.

% Utilization = ((weight of Iso-α-acids in wort or final beer)/ (weight of α-acids added to wort or beer)) × 100

Factors Affecting Utilization

Not all of the bitterness potential from the alpha-acid in the hop is utilized, which can be attributed to a number of reasons:

1. Form of Hops

The isomerization rate is initially affected by the form of the hops. Isomerization is slower and at a much lower rate with whole hops or plugs, slightly faster with standard pellets, and greatest with extracts.

2. Boil Conditions

Boil conditions can affect isomerization in a number of ways. For instance, the longer the boil continues, the more isomerization takes place, though eventually the reaction reverses itself, degrading the iso-alpha acids.

3. Hopping Rate

Isomerization is also affected by hopping rate; as the hopping rate increases, the rate of isomerization decreases. This effect can be partially offset by adding bittering hops in stages.

4. Fermentation Conditions

Fermentation conditions can affect the amount of iso-alpha acids that remain in the beer in a number of ways. Loss of iso-alpha acids also occurs during fermentation as they are adsorbed onto the yeast cell walls.

5. Maturation and Filtration Conditions

After fermentation, maturation and filtration conditions affect the extent to which not only bitterness, but also other hops components survive in the finished beer.

Dry Hopping—Basics and Techniques

A stagnating beer market provides opportunities to arouse consumer interest in beer by promoting 'specialties' that focus on completely new variations in flavor and, in particular, new aromas.

Definition of dry hopping

Hops and hop products are usually added in the brewhouse. To maximize bitter substance utilization, early hop addition to the kettle is essential. However, this, in turn, evaporates aroma substances after only a few minutes of exposure to the boiling wort. Late hopping in the hot section results in a decent hoppy note, but hardly reproduces the character of the hop variety/varieties used.

This is due to the fact that the volatile aroma substances that characterize a specific hop variety are lost, even if only briefly exposed to high temperatures. Hopping in the cold section is an ideal way to impart an aroma note typical of the variety used. In this case the hops need to be added either during fermentation or to the storage tank.

Techniques

The addition of hops pellets to the storage tank, which can hamper clarification and ensuing filtration due to the presence of insoluble hop components in the green beer, is a drawback.

• *Tentative Solution*

The above problem can be minimized by putting the hops pellets into small sacks. The sacks need to be made of very fine mesh, especially if using pellets. However, this method does not provide optimum extraction of the aroma substances.

In a few cases, hops are added during fermentation. This usually takes place towards the end of fermentation with a view to using the yeast to reduce some of the oxygen introduced by the hops.

On the other hand, the CO_2 resulting from fermentation acts as a kind of carrier gas, eliminating some of the highly volatile hop oils that would be of particular significance for dry hop aroma.

The best extraction of hop aroma can be achieved by pre-dissolving the hop pellets. This can be done either with beer or water, which must be oxygen-free and fully demineralized. A weight-based ratio of pellets to beer (or water) of 1:20 is recommended in order to make mixing easier. The suspension can be added to the storage tank manually, but it can also be pumped in. Any oxygen intake must be avoided.

• *Possible Problems*

Depending on the process that is implemented, aroma substance contents can vary from one beer batch to another. It is therefore essential to focus on the type of addition, the contact time, the crop year and so on. An additional important issue is the danger of infection when hopping in the cold section, although, as far as gram positive bacteria are concerned, beer is protected by the bacteriostatic effect of hops.

To Control Brewing Hop Consumption:

The following points have to be considered in order to control the cost of hops:

• The initial cost of the hop compounds
• Transport and storage costs
• Loss of alpha-acid
• The utilization of the hop product in the beer

Calculations of Hop Dosing

$$\text{Kg Hop per brew:} \quad \frac{\text{Targeted BU} \times \text{Brew length (hl)}}{\text{\% Utilization} \times \text{\% } \alpha\text{-acids in hop used (kg)}}$$

Beer with a Hop Flavor

For example, a beer with about twenty to forty European Brewery Convention (EBC) bitterness units possessing a noticeable nasal hop aroma will be described. The original extracts of the bottom fermenting beer are about ten to twelve percent. The beer should have a noticeable hop flavor. This should be perceived as a pleasant hop note during smelling and tasting. The following application examples can be recommended:

- About thirty to forty bitterness units with beers from hundred percent malt, as the flavoring matrix is very intensive due to the exclusive use of malt
- About twenty to thirty bitterness units with the use of adjuncts, as the flavoring matrix is not as pronounced and, hence, the bitterness is perceived sensorically more strongly
- The first hop addition at, but not before, the beginning of the boil to increase the yield of bitter resins
- Preferably, part of the first hops addition as aroma hops, as this gives a mild bitterness to the drink
- Alternatively, as a first addition, only bittering hops proven of value for one's own beer varieties in brewing experiments
- Aromatization by a second hop dosage which, if possible, should not be boiled anymore—thus, the steaming-out loss is minimized; this guarantees maximum aroma transfer from the raw material hop to the beer
- The second hop dosage can occur at the end of the boil in the wort kettle or as a dosage in the whirlpool
- The aroma transfer is favored by cooling the wort before addition to the whirlpool 'pre-cooling' and thus minimizing the steaming-out loss
- Brewing experiments need to be carried out to make the decisions about which aroma hop variety is best suited; depending on the variety, there are different influences on the flavor and the aroma of the beer
- The intensity and the quality of the hop flavor can be individually optimized with the point of addition; the time frame for the last hop addition starts about ten minutes before the end of the boil and spans to an underback in the whirlpool

- The contribution to the taste of the beer increases as more of the hop matrix (such as pellets type 90) is charged—this is applicable for aroma as well as bittering hops
- Addition of iso-pellets of aroma hops at the end of the boil increases the flavor and bitter resins yield
- Addition of isomerized kettle extract during wort boiling increases the bitter resins yield and microbiological stability
- Iso-extract before filtration to correct the bitters
- Hop oil or 'hop oil fractions' before filtration for flavoring
- Reduced iso-extracts to stabilize the foam (tetrahydroiso - or hexahydroiso - α -acids products)
- Combinations of the mentioned products

The dosage quantity of aroma hops in breweries is mostly determined by the concentration of α-acids. More logical is the dosage depending on the concentration of the hop oil or, even better, depending on the linalool concentration or depending on the leading component of the hop aroma, thus ensuring that the same amounts of aroma substances are always added. Time-dependent fluctuation in the oil concentration can be adjusted. In particular, with subsequent aroma dosage, the α-acid dosage is irrelevant as only a very small part isomerizes. A goal of hop addition is maximum aroma transfer. Linalool was found as the character impact compound for hop aromatic beers. The taste threshold value in beer is 2-80 μ g/l. A sensoric noticeable hop aroma in beer can be found at a linalool concentration of about 20 μ g in lager beers. This value depends on the particular taste profile of the beer. The hop aroma can be described as 'flowery perfumed' from about 50 μ g/l. Apart from linalool, other hop aroma substances are involved in the hop aroma. These are in parts below the taste threshold value of beer, but can contribute to the taste profile of the beer via synergistic effects. The taste stability of the beer can be positively influenced by late aroma addition. The stale flavor of beer is masked by a discreet but noticeable hop aroma.

Microbiology

Hops increase the microbiological stability of beer. Non-isomerized α-acids have three to four times stronger antimicrobial effect on beer-damaging microorganisms of the species *Lactobacillus* and *Pediococcus* than iso-α-acids. With late hop addition, the concentration of α-acids can be increased in the beer. Similar to foam, increasing hydration of the iso-α-acids has a positive effect on the microbiological stability of the beer.

Hop bitter resins can be listed beginning with the worst to the best microbiological beer stability as follows: un-hopped beer <iso-α-acids <rho-iso-α-acids <tetrahydroiso - α - acids < 50% tetrahydroiso - α - acids + plus 50% hexahydroiso-α - acids < α-acids.

During the operation of a biological acidification, facility attention should be paid such that no hop bitter resins comes into contact with the lactobacillus cultures (*Lactobacillus amylolyticus*), as these do not possess hop tolerance. For the biological acidification facility, re-feeding of the hot trub into the Lauter Tun may be started only after the first wort is withdrawn.

5

Water

Introduction

Breweries use large amounts of water (generally referred to as 'liquor' in the United Kingdom). As beers usually have water content of ninety-one to ninety-eight percent, and the amounts lost due to evaporation and with by-products are relatively small, it follows that a large volume of wastewater is produced. Sometimes, this is also a result of operational inefficiencies. However, it is to be noted that breweries operate in efficient but different ways, and with different product ranges, and have substantially different water requirements. Apart from brewing, sparging and dilution liquors, water is used for a range of other purposes. These include cleaning the plant using manual or CIP systems, cooling, heating (either as hot water or after conversion into steam in a boiler), water to occupy the lines before and after running beer through them, for loading filter aids such as Kieselguhr, for washing yeast and for slurrying and conveying waste, as well as for washing beer containers such as returnable bottles and kegs. The acquisition and treatment of liquor and the disposal of the brewery effluents are expensive processes.

Historically, different regions became famous for particular types of beer and, in part, these beer types were defined by the water available for brewing. For example, Pilsen, which has very soft water, became famous for its pale and delicate lagers. Burton-on-Trent, with its extremely hard water, rich in calcium sulfate, is famous for its pale ales, while Munich is well-known for its dark lagers, and Dublin (which has similar soft water) for its stouts. Breweries may receive water from different sources, which may be changed without warning. Salt content in water supplies may vary between day and night, from year to year and between seasons. It is now usual for breweries to adjust the composition of the water they use. In few regions, saline water must be used (even seawater). In principle, several desalination methods might be used, but in practice it seems that purified water is obtained from seawater either by a highly thermally efficient distillation, which is very costly, or by reverse osmosis. Usually, breweries obtain their water from their own wells, springs or boreholes (surface water is avoided where possible).

Most brewers find it necessary to treat the water coming into the brewery. A large variety of substances may be present in water, and different treatments are needed to deal with them. Different types of water require different treatments, and brewers require grades of water treated in different ways depending on the uses to which it will be put. In some instances, it may only be necessary to pre-treat the liquor, while extensive further treatment is needed for others.

As each water treatment technology removes only a specific type of contaminant, none can be relied upon to remove all contaminants to the levels required for critical applications. A well-designed water treatment system uses a combination of treatment technologies to achieve the required final water quality.

Each of the treatment technologies must be used in an appropriate sequence to optimize their particular removal capabilities. The schematic below shows a central laboratory water treatment system designed to produce water for critical applications.

The first step is use of pre-treatment equipment specifically designed to remove contaminants in the feedwater. Pre-treatment removes contaminants that may affect treatment equipment located downstream, especially ion exchange, and reverse osmosis (RO) systems. Examples of pre-treatment are:

- Particulate filters for sediment/silt/particulate removal
- Carbon filters (or tanks) for chlorine removal
- Softening agents to remove minerals that cause 'hard' water

The next treatment step is either ion exchange or Reverse Osmosis (RO). Reverse Osmosis removes ninety to ninety-nine percent of all contaminants found in water—hence, it is the heart of a well-designed water treatment system.

However, the tight porosity of the RO membrane limits its flow rate. Therefore, a storage container is used to collect water from the system and distribute it to other points-of-use, such as polishing systems.

Polishing systems purify pre-treated water, such as RO water, by removing trace levels of any residual contaminants. Treating raw tap water using such a system would quickly exhaust its capacity and affect final quality. Polishing elevates the quality of pre-treated water to 'ultra-pure' water.

A typical polishing system may consist of activated carbon, mixed-bed deionization, organic scavenging mixtures and 0.22 µm final filtration. Systems can also be enhanced with ultra filtration, ultraviolet oxidation or other features for use in specific applications.

This combination of treatment technologies, combined with proper pre-treatment, will produce water that is virtually free of ionic, organic and microbial contamination.

Water impacts beer in three ways:

1. Mashing process—impact on enzyme efficiency and flavor of wort
2. Hop utilization—water pH effect on hop utilization
3. Beer flavor—water adds flavor directly to the beer itself

The effect of brewing water on beer can be characterized by six main water ions: Carbonate, sodium, chloride, sulfate, calcium and magnesium.

One could get a water report from the local municipality/government-approved laboratory for the mineral content of the water to be used in the brewing process. In a water report, these are listed as parts per million (ppm), which is equivalent to one milligram per liter (mg/l). Each of the critical ions for lager beer type is described below:

- Soft water with low mineral content, low carbonates
- Calcium ions are important at levels > 30 mg/l

Brewing water can be adjusted by the addition of some brewing salts.

General Requirements

Water, in terms of quantity, is the most important raw material of beer. Therefore, the chemical and biological composition of water is significant, and there is no step in the brewing process that is not influenced by its constituents. Consequently, treatment of water is necessary in many cases. Since the breweries obtain their process water either from public mains, river and/or from their own wells, water treatment needs to be addressed in two aspects:

- Treatment of crude water to fulfill legal criteria
- Treatment of drinking water due to technological brewing requirements

Water used in the Brewery

The use of water in the brewery can be divided into three main categories:

1. Brewing water
2. Process water (for cleaning and sterilization)
3. Service water

Brewing Water

Brewing water plays a very important role in the flavor of the beer. Water makes up ninety-five percent of a beer's ingredients by weight (more than ninety-five percent), but the majority of brewers rarely consider what type of water to use. High consumption of good quality water is characteristic of beer brewing. Most brewers rarely give it much thought, and water is the least appreciated of beer ingredients. At the other extreme, many brewers spend hard-earned money on spring water because they don't fully understand how water affects the beer's flavor.

An efficient brewery will typically use between four and five liters of water to produce one liter of beer. Some (especially small) breweries use much more water. In addition to water used for beer mashing, wort boiling, sparging at Lauter Tun, beer or wort filtration, de-brewing, and bottle washing also use water for heating and cooling as well as cleaning out place (COP) and Cleaning in Place (CIP) of vessels, tanks and pipelines.

To know the character of the local available water source as well as how to adjust it to improve the beer is a critical skill, particularly for advanced brewers.

Table 5.1: Brewing Water Composition Requirements

Composition	Max. Level (mg/l)	Referred Concentration (mg/l)
Total Solids	500	-
Alkalinity (as $CaCO_3$)	50	0–25
Chloride	250–300	50–200
Sulfate	500	-
Nitrate	30	0–0.2
Fluoride	1.2	Not necessary
Silicates (as SiO_2)	10–50	0.1
Arsenic	0.05	-
Chromium	0.05	-
Copper	1.0	-
Lead	0.05	-
Manganese and iron	0.1	-
Phenolic substances	0.001	-
Selenium	0.01	-
Zinc	5	-
Oxidizable substances ($KMnO_4$)	< 100	-
Residual alkalinity	-	2–3°D hardness
Calcium (kettle and sparge)	-	40–70
Calcium (In wort boiling)	-	80–100
Calcium (remaining in beer)	-	60–80

Water for diluting beer (for high-gravity brewing) must correspond to the following average specifications:

- Dissolved oxygen content should be as low as possible (<0.05 ppm)
- The CO_2 content of the water should be equal to or slightly above the level of the beer to be diluted; attention should be paid to the O_2 content of the CO_2
- The water must be free of microorganisms

Process Water—for Cleaning and Sterilization

For cleaning and sterilization operations, it is necessary to pay great attention to the water quality. The water should be of potable standard and free from off-flavors for cleaning operations. For sterilization, the water must be free of microorganisms and off-flavors. For cleaning and rinsing, it is desirable to use potable water containing 0.1 mg/l of free chlorine at the point of use.

Attention must be paid to the possibility of interaction of phenolic compounds with chlorine with formation of chlorophenols. The flavors threshold of chlorophenols is very low. Water for dissolving detergents should be low. Water for disinfectants should be soft and sterile.

Service Water

Water for boilers must correspond to the specifications of the boiler IBR Act or the manufacturer of the boiler. Water for a refrigeration plant must also correspond to the specifications of the manufacturer of the compressors or refrigeration plant. Chemical and microbiological treatment of water is necessary.

Water used by a pasteurizer must be treated to reduce its content of minerals salts, and to prevent the growth of algae or microorganisms. The pH of the water must be constantly controlled. The water treatment must be compatible with the materials of the pasteurizer so as to avoid any corrosion.

Brewery Water Consumption

On the basis of water usage ratio of 6 hl/hl of beer, the following typical consumption proportions can be made:

- Brewing Water : 45%
- Process Water : 52%
  (Water for cleaning and sterilization)
- Service Water : 3%

Of the 6 hl of water consumed per hl of beer, following are the estimated break-up of the usage/loss of water at various stages in the brewery along with the generation of effluents:

- In finished beer : 15%
- Loss by evaporation : 3%
- Loss with spent grain and spent yeast : 2.5%
- Effluent : 60.5%
  (Brewhouse, fermentation, lagering, filtration, bottle washer, filler, pasteurizer, cleaning and CIP, etc.)
- Effluent—domestic : 2%
- Effluent—water treatment : 17%

Water Treatment

There are a number of chemical and physical alternatives available for water treatment. In the scope of this chapter, only the most common methods will be discussed. In recent years, the following compounds have had to be disposed of as the most common problematic materials: Iron, manganese, nitrates, halogenated hydrocarbons and pesticides.

Table 5.2: Quality criteria of brewing water

Characteristic	Value	Reason
pH	7–8	Too acidic: danger of corrosion Too basic: inhibition of enzymes
P	0–0.3 mval/l	Water does not aggression CO_2, but only low fraction of CO_3^{2-} and OH^- ions
M	0.7–1.2 mval/l	Only low residual of acid destroying HCO_3^-; low fraction but positive for palatable taste
Non-carbonate Hardness	At least twice, better three times the carbonate hardness	Balanced alkalinity
Residual alkalinity	-2 to 2°dH < 5°dH < 10°dH	For Pilsner beers For light beers For dark beers
Sulfate	100–150 mg/l	Dry bitterness, tendency to a hop aroma
Chloride	< 150 mg/l 300 mg/l	Salty taste, corrosion Toxic to yeast
Nitrate	< 25 mg/l	Fermentation disturbances are avoided; low value is better as nitrate is also introduced into the beer by hop and malt
Sodium	30–60 mg/l	Sour and salty taste

Characteristic	Value	Reason
Potassium	300–500 mg/l	Saline effect
Iron	< 0.1 mg/l	Flaws in taste, danger of gushing, danger of turbidity, beer taste in stability, brown color to the foam
Zinc	0.08–0.20 mg/l 0.6 mg/l	Activates fermentation and limits formation of H_2S Inhibits yeast growth and fermentation
Copper	> 10 mg/l	Toxic to yeast
Manganese	< 0.2 mg/l	Affect colloidal stability
Free aggressive CO_2	-	Danger of corrosion

Removal of Problematic Organic Substances

Deferrization and Demanganization

This process is necessary—even from a brewing technological point of view—if the iron concentration is above 0.12 mg/l and the manganese concentration is above 0.05 mg/l. Through oxidation with adequate aeration devices, water-insoluble compounds are formed and removed from the water by filtration.

Through aeration, the soluble iron (2^+) compounds are transferred to insoluble iron (3^+) compounds. Organically bound iron is precipitated by the addition of oxidants (e.g. cupper permanganate and flocculant Fe (2^+) salts). The reaction is pH dependent, where the oxidation rate increases with higher pH.

Demanganization is a considerably more sensitive process. Manganese (2^+) is oxidized to manganese (4^+). A precipitation of manganese oxide is achieved at a pH above 9 by the addition of oxidants (such as potassium permanganate). A biochemical oxidation by manganese bacteria can occur simultaneously on the filter bed. Consequently, a longer adjustment period during demanganization is to be expected as compared to the deferrization filter.

Two separated filter units are normally advantageous for the removal of iron and manganese. Deferrization occurs in the first filter (saving cleaning water, iron compounds are easily rinsed off).

In the second filter, demanganization is carried out (increased holding time, gradual coarsening of the filter material through black manganese dioxide).

Nitrate Reduction

Because of the intensive use of nitrogen fertilizers in agriculture, the nitrate level of ground waters has significantly increased. Chemical as well as biological methods are used for the removal of nitrate from crude water. The following methods of treatment are available:

Chemical Treatment:

- Ion exchanger
- Reverse osmosis
- Semi-permeable membranes
- Electrolysis

Biological Treatment:

- Heterotrophic denitrification
- Autotrophic methods

Deaeration

The high-gravity brewing procedure involves deaeration techniques for diluting water. Several techniques are available, such as:

- Thermal degassing
- Vapor stripping
- Vacuum degassing
- Combined systems
- CO_2 stripping

Volatile substances (such as halogenated hydrocarbons) can be expelled to a large extent by 'stripping' the water with air (for example, intensive aeration).

Dechlorination (Carbon Adsorption)

Water used in the brewing process should be from of chlorine. Chlorine can be adsorbed by active carbon.

Carbon adsorption is a widely used because of its ability to improve water by removing disagreeable tastes and odors, including objectionable chlorine. Activated carbon effectively removes many chemicals and gases; in some cases, it can be effective against microorganisms. However, generally, it will not affect total dissolved solids, hardness or heavy metals. Only a few carbon filter systems have been certified for the removal of lead, asbestos, cysts and coliform. There are two types of carbon filter systems, each with advantages and disadvantages: Granular activated carbon, and solid block carbon. These two methods can also work with an RO system.

The ability of an activated-carbon filter to remove certain microorganisms and certain organic chemicals, especially pesticides, THMs (the chlorine by-product), trichloroethylene (TCE), and PCBs, depends upon several factors, such as the type of carbon and the amount used, the design of the filter and the rate of water flow, how long the filter has been in use, and the types of impurities the filter has previously removed.

The carbon adsorption process is controlled by the diameter of the pores in the carbon filter and by the diffusion rate of organic molecules through the pores. The rate of adsorption is a function of the molecular weight and the molecular size of the organics. Certain granular carbons effectively remove chloramines. Carbon also removes free chlorine and protects other treatment media in the system that may be sensitive to an oxidant such as chlorine.

Carbon is usually used in combination with other treatment processes. The placement of carbon in relation to other components is an important consideration in the design of a water treatment system.

Advantages:

- Removes dissolved organics and chlorine effectively
- Long life (high capacity)

Disadvantages:

- Can generate carbon fines

Activated-carbon Filtration (ACF)

Activated-carbon filtration is useful for polishing and removal of problematic substances in the water. There, contaminants adsorb onto the surface of the activated carbon and are thus fixed. In this repeatedly recommended and used process, the problem is shifted as this alternative provides no elimination and mineralization of these problematic substances (see section *Combination Processes Using Oxidation/UV Irradiation*) The ACF helps, among other things, the removal of:

- Odorous substances and flavorings
- Discoloration

- Organic water constituents (such as pesticides)
- Halogenated hydrocarbons
- Chlorine and chlorine derivatives
- Microorganisms

In order to enable sterilization of the filters and thus avoid microbiological contamination, the ACF should be made of stainless steel. Active carbon is widely used due to its large specific surface area ($700–1500$ m^2/g).

Combination Processes Using Oxidation/UV Irradiation

Both sterilization and mineralization of organic contaminants in the water are possible through the combination of chemical and physical reaction mechanisms. The chemical oxidation occurs with hydrogen peroxide or ozone. Thin-walled radiators with high application rates are used in the subsequent UV irradiation.

Common Processes for Brewing Water Treatment

The composition of the brewing water is of great importance for the quality as well as the character of the beers. Apart from the change in hardness former (softening), the reduction of nitrate concentration in the brew water plays an ever-increasing role.

Treatment with Lime

Lime treatment precipitates alkaline-earth bicarbonates by the following reactions:

$$Ca(HCO_3)_2 + Ca(OH)_2 \rightarrow CaCO_3 + 2H_2O$$
$$Mg(HCO_3)_2 + Ca(OH)_2 \rightarrow CaCO_3 + MgCO_3 + 2H_2O$$
$$2NaHCO_3 + Ca(OH)_2 \rightarrow Na_2CO_3 + CaCO_3 + 2H_2O$$

Carbon dioxide is removed by the following reactions:
$$CO_2 + Ca(OH)_2 \rightarrow CaCO_3 + H_2O$$

The efficiency of this softening process is influenced by the following factors:

- High temperatures: Above $12°C$
- Strong convection: Stirring, pumping, fluidized-bed reactor
- Crystallization seeds: Contact sludge, sand
- Pressure or rapid decarbonization

The pH of the brew water after softening is relatively high at about 8.5. This has no influence on the mash pH as the buffering capacity of the water is low. To control the quality of the brew water, the *p* and *m* values are determined. The *m* value serves to determine the softening effect; the *p* value gives an indication of the correct dosage of lime.

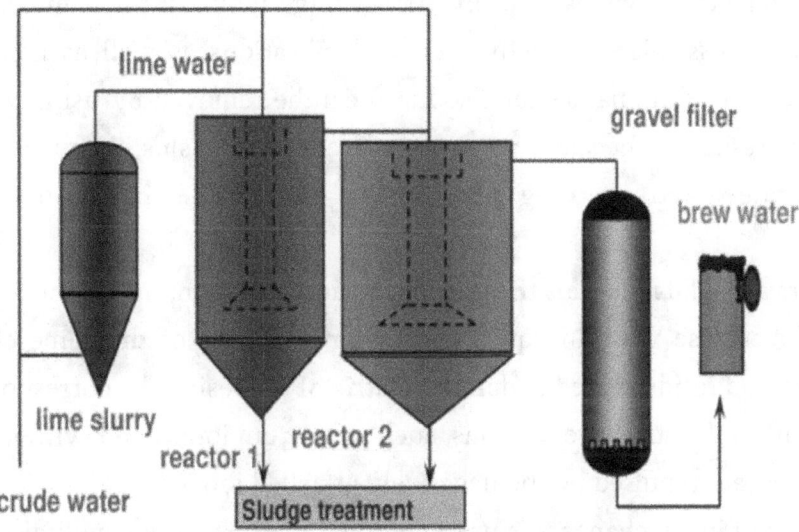

Figure 5.1: Lime precipitation unit

Kolbach (1953) indicated that different treatments were necessary for the following groups of water:

1. When the calcium hardness > total alkalinity (water rich in $CaSO_4$ or permanent hardness), the water is very easy to decarbonate
2. When the calcium hardness = total alkalinity (water that contains pure calcium carbonate), the water is very easy to decarbonate
3. When the calcium hardness < total alkalinity (water that contains, in addition to calcium bicarbonate, magnesium and sodium bicarbonate), the water is difficult to decarbonate

Treatment with lime may be difficult if water contains organic matter and magnesium salts.

Split Treatment

Magnesium bicarbonate can be removed by an excess of lime, according to the following reaction:

$$Mg(HCO_3)_2 + 2Ca(OH)_2 \rightarrow 2CaCO_3 + Mg(OH)_2 + 2H_2O$$

A large excess of lime is necessary (pH 10.5 – 11.0) for precipitation. After treatment the water must be neutralized with acid.

Lime treatment and split treatment methods of water decarbonization are large producers of effluent because of the volume of carbonate deposits.

Ion Exchange

The use of ion exchangers is widespread in the beverage industry for demineralization of crude waters, since this process allows selective removal of cations as well as anions. The hardness former of the water (calcium, magnesium, sodium) can be removed by using weak and/or strong acidic ion exchange resins. In the subsequent use of weak basic resins, the water can be completely demineralized. Nitrate-selective (strong basic) resins, on the other hand, only reduce the nitrate concentration.

Ions exchangers are solid materials that are able to adsorb cations and anions from an electrolytic solution, and replace these with an equivalent of amount ions of the same charge. In a cation exchanger, sulfonic acid residues are fixed in the matrix of the resin. The corresponding sodium ions can move freely within the structure. Cations, such as calcium ions, enter without problem; anions, such as chloride ions, are repulsed by the negatively attached ions.

A differentiation of ion exchangers is presented in Table 5.3. Only legally licensed regenerating materials can be used to regenerate the ion exchanger. Furthermore, the resins need to be safe for all foods—no flavorful, olfactorial or health-damaging materials should be emitted/given off.

Table 5.3: Differentiation of Ion Exchangers

Type	Removal of	Anchoring Groups	Counter Ion	Regenerant: Aqueous Solution of
Cation Exchanger				
Strong acidic	Ca, Mg, Na	Cl, SO_4, NO_3 HCO_3	H^+ Na^+	HCl, H_2SO_4 $NaCl$
Weak acidic	Ca, Mg	HCO_3, CO_3	H^+ H^+ Na^+	HCl, H_2SO_4 CO_2 $NaOH$
Anion Exchanger				
Strong basic	All anions Cl, SO_4, NO_3, HCO_3	All cations Na, K, Ca, Mg	Cl^- SO_4^{2-}	$NaCl$, HCl Na_2SO_4, H_2SO_4

Weak basic	Anions of strong acids Cl, SO₄, NO₃	All cations Na, K, Ca, Mg	Cl⁻ OH⁻ (free basic form) HCO₃	NaCl NaOH CO₂

The following points should be considered for the planning of an ion exchanger plant:

- Free from floating materials
- Polishing filter in the case of organically polluted waters
- Deferrization and demanganization
- Free of chlorine and chlorine derivatives
- Activated-carbon filter prior to softening to protect the ion exchange material

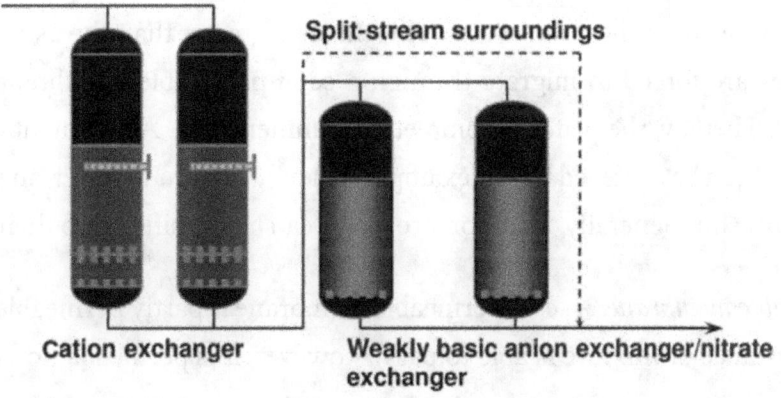

Figure 5.2: Schematic presentation of an ion exchanger

A common process combination of a cation and anion exchanger is presented in Figure 5.2. The removal of the former's hardness is carried out through cation exchange. The optimal following anion exchanger can be designed either to serve as a weak basic resin for complete demineralization or as a strong basic resin to selectively remove nitrate. Depending on the choice of the combination processes, a special regeneration scheme is applied.

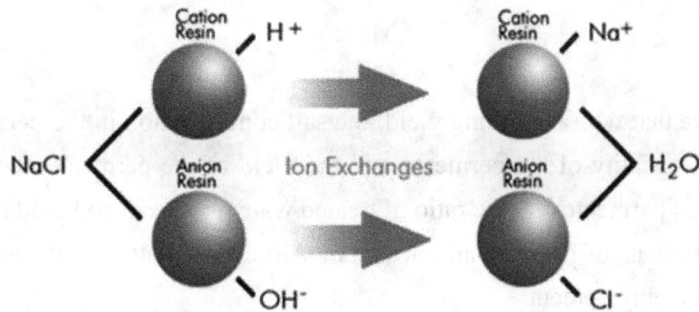

<u>Advantages:</u>

- Removes dissolved inorganic elements effectively
- Regenerable (service deionization)
- Relatively inexpensive initial capital investment

<u>Disadvantages:</u>

- Does not effectively remove particles, pyrogens or bacteria
- DI beds can generate resin particles and culture bacteria
- High operating costs over the long term

Membrane Processes

Reverse osmosis is normally used for treating water that is used in brewing. In this process, high pressure is administrated onto the crude water. This must be greater than the osmotic pressure, so that the water molecules are forced to migrate through a semi-permeable membrane, whereas the salts are mostly retained. Thereby, the water is completely demineralized. Adjustment of the desired water quality (see Table 5.2) can be achieved by, for example, a blend with crude water, an increase of hardness with lime slurry and so on. Generally, filtration processes can be classified into their separation spectra.

- *Semi-permeable membrane:* A semi-permeable membrane is partly permeable, (i.e.) only a certain part of a substance or mixture is able to pass); however, it is permeable towards the solvent
- *Permeate:* The permeate is the cleaned solution of the membrane process, (i.e.) the part of the feeding water that migrates through the membrane)
- *Concentrate (retentate):* In reverse osmosis, the concentrate is the concentrated water that, corresponding to the volume of obtained permeate, contains the dissolved substances of the feed in concentrated form
- *Yield:* The yield is the rate of permeate volume to the feed volume:

$$\text{Yield (\%)} = \frac{\text{Permeate}}{\text{Crude water}} \times 100$$

It is important to note that with increasing yield, the salt concentration in the permeate also increases. An optimum between the quality of the permeate and the yield of the permeate needs to be established for every use. The yield of permeate (i.e. the ratio of treated water to introduced crude water) can be arbitrarily adjusted through variations of pressure and flow rate. A brewing water treatment plant commonly works with a yield of about eighty percent.

For an economical operation mode of membranes, there are a number of limiting factors, which could either damage the membranes or lead to a blockage. For example, the barium concentration of the crude water should be below 0.1 mg/l to avoid irreversible precipitation of barium sulfate onto the membrane.

Advantages:

- Effectively removes all types of contaminants to some extent (particles, pyrogens, microorganisms, colloids and dissolved inorganic elements)
- Requires minimal maintenance

Disadvantages:

- Flow rates are usually limited to a certain m³/day rating

Microbiological Control (Sterilization of Water)

All types of water (liquors) contacting the product stream at any stage of the brewing process should be sterile—this is especially import with regard to water-borne organisms.

Bacterial contamination of water can occur at any stage during storage, even at high temperatures. For example, thermophilic bacteria (such as *bacillus stearothermophilus*) may survive in a hot brewing water tank at 80°C. A small leak of water across the paraflow heat exchange plates into the water stream can provide the bacteria with sufficient nutrient to grow to considerable numbers within hot water and can actually cause wort and beer clarity problems when the contaminated water is used for wort production.

Microbiological purity in the brewery is important—from wort cooling to the finished product. To make microorganisms inactive, the following techniques are used:

a. Chlorination:

Chlorination is carried out by the application of chlorine gas, hypochlorous acid, hypochlorite solution, or chloramines. Before application, a water analysis should be available to guide the correct choice of treatment. It is important to determine the free chlorine concentration at the point of water utilization.

b. Ozonization:

This treatment is very powerful and inactivates microorganisms a thousand times more rapidly than chlorine. Ozone, as well as disinfecting, degrades organic substances, which are flocculated. The amount of ozone to be injected depends on the water composition and varies between 0.5–3.0 ppm. Precautions concerning corrosion must be taken, as ozone is a powerful oxidant.

c. Ultraviolet Radiation:

Ultraviolet radiation for water disinfection using mercury vapor lamps at wavelengths between 200 and 300 nm are efficient for destroying microorganisms. It is important to check water composition for concentrations of organic substances and microorganisms.

Ultraviolet radiation has widely been used as a germicidal treatment for water. Mercury low-pressure lamps generating 254 nm UV light are an effective means of sanitizing water. The adsorption of UV light by the DNA and proteins in the microbial cell results in the inactivation of the microorganism.

Recent advances in UV lamp technology have resulted in the production of special lamps that generate both 185 nm and 254 nm UV light. This combination of wavelengths is necessary for photo oxidation of organic compounds. With these special lamps, Total Organic Carbon (TOC) levels in high purity water can be reduced to 5 ppb.

Advantages:

- Effective sanitizing treatment
- Oxidation of organic compounds (185 nm and 254 nm) to < 5 ppb TOC

Disadvantages:

- Decreases resistivity
- Will not remove particles, colloids, or ions

d. Treatment with Silver:

This electrolytic process uses a silver anode and a carbon or stainless steel cathode. The amount of free silver needed depends on the water's composition and varies between 0.05 and 0.2 mg/l for protection against infection; for disinfection, the silver concentration is increased to 0.5—1.0 mg/l.

Conclusions

It is believed that both ion exchange and RO technologies will continue to grow strongly in the near future. The RO membrane process clearly has a bright future. In addition, with water treatment processes demanding greater efficiency and selectively, ion exchange also has a bright future. The modern technology of water treatment requires the combination of several technologies. For example, the integration of different membrane processes, such as ultra filtration and RO membrane separation, or the combination of several techniques such as RO and ion exchange.

The two state-of-the-art process solutions (RO and ion exchange) allow the process designer to optimize the water treatment process with the goal of a lower cost and higher degree of reliability.

6

Yeast

TAXONOMICALLY, THE TWO species, *Saccharomyces uvarum* (carlsbergensis) and *Saccharomyces cerevisiae*, have been distinguished on the basis of their ability to ferment disaccharide melibiose. There is a problem in classifying strains in the brewing context—the taxonomist dismisses the minor difference between the strains encompassing brewing, baking, wine distilling and laboratory cultures, which are of great technical importance to the brewer.

Strains of *S. uvarum (carlsbergensis)* possess the MEL genes. They produce the extracellular enzymes alpha-galactosidase (melibiose) and are able to utilize melibiose, whereas strains of *S. cerevisiae* do not produce galactosidase and, therefore, are unable to utilize melibiose. In 1984, yeast taxonomists consolidated *S. uvarum (carlsbergensis)* and *S. cerevisiae* into one species, *S. cerevisiae.*

Pure Yeast Cultures

Introduction

Yeasts belonging to the genus Saccharomyces are often referred to as the oldest plants cultivated by man. The practice of using a pure yeast culture for brewing was started by Emil C. Hansen in the Carlsberg laboratories over a hundred years ago. By using dilution techniques, he was able to isolate single cells of brewing yeast. This allowed one to test and then select the specific yeast strains that gave the desired brewing properties. The first pure yeast culture was introduced into a Carlsberg brewery on a production scale in 1883 and the benefits of pure culture quickly became clear.

Strain Selection

In a laboratory or on a small pilot plant scale, it is not difficult to keep a culture pure and healthy. However, large-scale management of the yeast is a more difficult task once the culture enters the brewery environment. There are three basic needs:

1. A regular pure yeast culture source
2. Maintenance of the brewery yeast supply
3. Microbiological control

In many breweries, fresh yeast is propagated every eight to ten generations (fermentation cycles), or earlier if contamination or a fermentation problem is identified. The systematic use of clean, pure, and highly viable cells ensures that bacteria, wild yeast, or yeast mutants, such as respiratory deficiency petites, do not lead to inconsistent fermentations and off-flavor development.

Lager yeast is normally a pure culture, whereas ale yeasts have often been a mix of strains. The pure culture practice is invaluable in ensuring that any wild yeast are quickly detected and not allowed to proliferate into a significant problem for the brewer. Moreover, undesirable mutations of the parent strain, which often occur over long usage, are kept to a minimum. Today, various procedures are used to isolate pure cultures from pitching yeast including culturing from a single cell by Linder's hanging drop technique, micromanipulation, or culturing from a single colony isolate from an agar or gelatin plate.

Storage of Cultures

The most important consideration in the maintenance of a culture collection of brewing yeasts is that the stored cultures and their subsequent progeny continue to accurately represent the strains originally deposited. The yeast preservation method should confer maximum survival and stability and be appropriate to the laboratory facilities available. There are many methods available to store yeast and bacteria. The most common preservation methods currently in use are subculture, drying or desiccation, freeze drying, and freezing or cryopreservation.

Propagation and Scale-Up

The first yeast propagation plant was developed by Hansen and Kuhle and consisted of a steam-sterilizable wort receiver and propagation vessel equipped with a supply of sterile air and impeller. The basic principles of propagation devised by Hansen in 1890 have changed little. The propagation can be batch-wise or semi-continuous. There are usually three stainless steel vessels of increasing size equipped with attemperation control, sight glasses, and non-contaminating venting systems. They are equipped with a CIP system and often have in place heat sterilizing and cooling systems for both the equipment and the wort. The yeast propagation system is ideally located in a separate room from the fermenting area, with positive air pressure as well as humidity control and air sterilizing systems, disinfectant mats in doorways, and limited access by brewing staff.

During yeast propagation, the brewer wishes to obtain a maximum yield of yeast but also wishes to keep the flavor of the beer similar to a normal fermentation, so that it can be blended into the production stream. As a result, the propagation is often carried out at only a slightly increased temperature and with intermittent aeration to stimulate yeast growth. The propagation of the master culture to the plant fermentation scale is a progression of fermentations of increasing size (typically 4–10x), until enough yeast is grown to pitch a half size or full commercial size brew.

Wort sterility is normally achieved by boiling for thirty minutes, or the wort can be pasteurized using a plate heat exchanger and passed into a sterile vessel and then cooled. Wort gravities range from 10°Plato to 16°Plato. Depending on the yeast, zinc or a commercial yeast food can be added. Aeration is important for yeast growth and the wort is aerated using oxygen or sterile air and antifoam is added if necessary. Agitation is not normally necessary as the aeration process and CO_2 evolved during active fermentation are sufficient to keep the yeast in suspension.

The exact details of the yeast propagation will depend on whether it is a small brewery or a larger brewery utilizing high-gravity fermentation and depending on the propagation equipment available. The steps can be larger and the temperature varied from 12–25°C with resultant longer propagation times at the lower temperature. Scale-up steps are kept small at the early stages to ensure good growth.

Yeast Selection

Specific selection of appropriate brewing yeast for the specific process is mainly carried out through comparison of specific requirements with the information contained about the individual strains of yeast. In order to match the requirement of the brewery in the best possible way, the following information is required:

- Raw materials to be used
- Wort specifications
- Yeast characters required
- Beer profile required
- Fermentation characteristics

Single Cell Culturing

A yeast sample may consist of several strains of yeast. Single cell pure culturing is carried out to ensure hundred percent pure yeast. The single cell culturing includes the following steps:

- Transfer of single cells to moisture chamber
- Propagation of 2×20 yeast cells in wort drops for approximately two days
- Transfer of 10 single cell suspensions to 10 ml wort

Yeast Strain Improvement

Using the single cell culturing and yeast typing methodology, improvements in the brewery's existing yeast strain performance is a possible option to replacing the strain. By screening individual cells taken from the brewery's culture for specific technical or aromatic properties (for example, improved flocculation or aromatic properties), improvement in yeast performance can be obtained.

Many breweries has seen that this is a more acceptable way of improving yeast performance and beer quality compared to the introduction of new strain.

Brewing Yeast Nutritional Requirements

The brewing yeast will be grown successfully, with good vitality, if there is adequate supply of nutrients, fermentable sugar, amino acid, vitamins and minerals. These nutrients are always found enough in wort that it is not limiting the growth of yeast.

1. Carbohydrates (Carbon source)

Low molecular weight sugars such as glucose, fructose, maltose, sucrose and maltotriose are available for yeast growth. The bigger carbohydrates (such as dextrose) are not used by the yeast cell. The wort consists of about eighty percent of fermentable extract. The balance twenty percent consists of non-fermentable products such as dextrose, β-glucans, pentosans, and oligosaccharides.

2. Amino Acid (Nitrogen source)

The yeast cells use amino acid to synthesize more amino acids and, in turn, to synthesize proteins. Amino acids, collectively referred to as 'free amino nitrogen (FAN)', are the principal nitrogen source in wort and are an essential component of yeast nutrition.

In brewing yeasts growing on wort, the uptake of amino acids is an ordered process. Pierce (1987) divided amino acids into four classes based on the order of assimilation from wort during fermentation:

Class **A:** This group will be the fastest taken by the yeast cell in the beginning of fermentation, but synthesized later (Example: Aspartic acid, Asparagin, Glutamic acid, Glutamine, Threonine, Serine, Methionine).

Class **B:** The yeast cell will be synthesized at early stages of fermentation, but prevented later (Example:. Isoleucine, Valine, Phenylalanine, Glycine, Alanine and Tyrosine).

Class **C:** These amino acids are the yeast cells taken only from wort, it cannot be synthesized it owns (Example: Lysine, Histidine, Arginine and Leucine).

Class D: This amino acid is not utilized during beer fermentation (Example: Proline).

3. Vitamins

Vitamins such as choline, thiamin (B1), folic acid, nicotinic acid, calcium pantothenate, pyridoxin and biotin are essential for enzyme function and yeast growth. These vitamins are always sufficiently found in wort that it is not limiting the growth of yeast. In wine and cider production, one may have to add salts that contain nitrogen (such as ammonium phosphate), and sometimes even thiamine and riboflavin.

4. Minerals

Yeast cells cannot grow unless they are provided with a source of essential minerals. These include phosphate (P), potassium (K), calcium (Ca), magnesium (Mg), and sulfur (S) and trace elements. For example, P is involved in energy conservation, is necessary for rapid yeast growth, and is part of many organic compounds in the yeast cell.

5. Zinc

Zinc (Zn) is the most important trace element; the optimum content should be 0.15 ppm. Need of Zn depends largely on the yeast strain—some yeast strains may need more zinc and, therefore, some breweries add it to the wort. Zinc assists in protein synthesis in yeast cells and controls their nucleic acid and carbohydrate metabolism. It very important for stabilization of proteins and membrane systems, activity of enzymes, protection of enzymes against denaturation, speeding up riboflavin synthesis, simulation of sugar uptake. Zinc may reduce enzyme activity in mashing. To increase Zn level, the brewer can theoretically select proper malt, higher malt modification, lower mash pH, lower mashing-in temperatures, shorter mashing times, use low mash concentrations and use some amount of husk or spent grain extracts.

Yeast Autolysis

Autolysis of the yeast cell is the break-up of the cell on its own. Deficiency of nutrients, high temperature and alkaline conditions can enhance autolysis. If yeast is left in the fully fermented beer for far too long, autolysis will occur and cause off-flavor in the beer. When a yeast cell dies, the reductase enzyme will rupture the cell wall's protein, thus releasing several off-flavors into the beer.

In biology, autolysis, more commonly known as self-digestion, refers to the destruction of a cell through the action of its own enzymes. It may also refer to the digestion of an enzyme by another molecule of the same enzyme.

The Effect of Yeast Autolysis

- Yeasty flavor
- Creosote-like foreign taste
- Increase of beer pH—release of amino acids
- Change in beer color
- Decreased biological and chemical stability
- Poor head retention (foam stability)
- The bitterness is broader and more persistent
- Lower diacetyl removal rate

- Beer infection occur more frequently
- Decrease flavor stability
- May cause of filtration problem

Prevention of Yeast Autolysis

- Always keep beer and yeast cool
- Harvesting all the yeast from the beer after the end of fermentation, inasmuch as possible
- Minimize top pressure
- Cropping dead cell at the beginning of fermentation (normally after twenty-four hours)
- Cropping dead cell before end of fermentation
- Ensure to re-pitching with high viability yeast (dead cells should not be more than fifteen percent)

Yeast Cleaning

Yeast infections in beer, which are often referred to as wort bacteria, lactic acid bacteria and wild yeast infections, cause off-flavor, haze and a short shelf life in beer. So, pitching yeast must be always cleaned. These are three techniques that can be used for cleaning brewing yeast before re-pitching it into the wort.

1. Mechanical cleaning
- Screen cleaning—to remove cold break, collapsed foam, risk of contamination and loss of yeast
- Wash with cold sterile water to remove bacteria and broken cells—it is the period where the individual bacteria are maturing and not yet able to divide; during the lag phase of the bacterial growth cycle, synthesis of RNA, enzymes and other molecules occurs
2. Chemical cleaning
- Acid wash—as brewing yeast has more tolerance to low pH than bacteria, addition of acid to a pH range 2.1–2.5 is enough to kill the infected bacteria

Phosphoric and citric acids offer the advantage of being weak acids and the yeast pH is more easily controlled, whereas with strong acids (such as sulfuric acid) there are special handling procedures that have to be followed by operators. A slight overdose of strong acid will yield excessively low pH values.

Recommended technique to be followed for acid-washing of yeast:
- Use food grade phosphoric acid
- Wash the yeast as a beer or water slurry
- Chill both the yeast slurry and the acid to less than 5˚C
- Stir continuously and slowly while adding the acid to the yeast

- If possible, stir throughout the wash
- Never let the temperature exceed 5°C during the wash
- Check the pH of the slurry
- Do not wash for more than two hours
- Pitch yeast immediately after washing
- Do not wash unhealthy yeast or yeast from the fermentations with greater than eight percent ethanol present

Need to Evaluate Pitching Yeast

There are five reasons behind evaluating pitching yeast:

1. If you have handled your yeast correctly, they could produce high quality products in a cost-effective manner, despite the stresses that are clearly evident in modern breweries' environments (such as alcohol, CO_2 toxicity, nutrient limitation and cold shock)
2. It is beneficial for brewers who need to monitor not only yeast performance but aspects of the production process that can influence this performance as well
3. Because it is imperative to measure and evaluate pitching yeast quality
4. We do not have all details of yeast physiology, but have a number of indicative, analytical tools to measure pitching yeast quality
5. To overcome deterioration of yeast performance

Yeast Development

- Yeast harvest should be minimum 1.5 times pitching amount
- Maximum yeast concentration should be three to four times the amount of pitching yeast
- Too high yeast production leads to increased beer loss, lower yield of bitter substances, decreased ester formation
- Risk of off-flavors (higher alcohols)
- Too low yeast production leads to slower fermentation and problem in yeast supply

Yeast Viability and Yeast Vitality

Yeast Viability vs. Yeast Vitality

Yeast viability is a measure of the number of living cells, which is a measure of yeast's ability to ferment—a property not possessed by dead cells.

Yeast vitality is a measure of yeast activity or fermentation performance. The vitality of yeast can also be expressed as a function of the total cell viability and the physiological state of the viable cell population.

Yeast Viability Testing

The reference analysis for viability is the plate count measurement. The most commonly used method to determine the viability of the pitching yeast is the slide culture technique done by trained microbiologists. A suitably diluted suspension of yeast is applied to a microscope slide and covered with a thin layer of nutrient media. A sterile cover slip is positioned over the yeast and the slide is incubated for no longer than eighteen hours at room temperature. The slides that give rise to micro colonies are viable. Single cells that do give rise to micro colonies are scored as dead. The percentage of viable cells is called the 'budding index'.

Yeast Vitality Testing

Yeast vitality can be measured using the Acidification Power Test, which measures the ability of the yeast to acidify a basal medium in the presence or absence of sugar and, thus, reportedly tests both the membrane functionality and the sugar catabolic rate of the culture. However, this test is not rapid enough to allow its use on all pitching yeast samples on a routine basis.

Note:
Normally, only the brew master determines how many dead cells are there in the pitching yeast in the yeast storage tank ahead of the next use of the pitching yeast.

Yeast Pitching Rate

Yeast pitching is decided by a number of factors:
- Wort gravity
- Wort constituents
- Temperature
- Degree of wort aeration
- Previous history of the yeast

Ideally, it is desirable to have a minimum lag in order to obtain a rapid start to fermentation, which then results in a fast pH drop, and ultimately assists in the suppression of bacterial growth. Generally, ten million cells/ml is considered an optimum level of pitching rate and results in lager yeast reproducing four to five times.

Yeast Growth & Final Beer Character

In case of higher yeast growth, will have a major influence on flavor and aroma compounds such as fusel alcohol, esters, and diacetyl (VDKs). The concentration of VDKs is related to yeast growth and accelerated fermentation may result in higher VDKs than traditional fermentation because of greater precursor production or slower conversion to VDKs.

Increased pitching rate will increase the loss of bitter substances and polyphenoles because of adsorption of these substances to the increased cell surface area due to the many new cells formed.

Yeast Storage

Ideally, yeast is stored in a room designed to be easily sanitized, containing sufficient supply of sterile water and a separate filtered air supply under positive pressure to prevent the entry of contaminants, and having a temperature of 0°C. Alternatively, insulated tanks in a dehumidified room are employed. Removal of unnecessary equipment and tools from the room has proven beneficial.

Yeast is most commonly stored under six inches of beer or under water or two percent potassium dihydrogen phosphate solution. Reduction of available oxygen is important during storage and minimum yeast surface areas exposed to air is desirable.

Shipping of Yeast

The pure culture is often maintained and supplied from a commercial laboratory and sent through the mail growing on a slope or in a pressed form or the brewery can supply a slope from its own central laboratory to branch plants. More commonly, if a brewery has a number of branch plants, a pure yeast culture system is situated in only one plant and this plant supplies the other branch plants with the yeast needed to pitch a larger fermentation vessel. For breweries with multiple plants, inter-plant homogeneity of the culture yeast is always a concern when consistency of flavor between plants is critical. Sending yeast from a central culture plant in sufficient quantity ensures conformity between plants.

General principles to be observed when shipping yeast from a central brewery or culture plant to other breweries:

1. Selecting yeast with the highest viability

2. Absence of contaminants

3. Removal of fermentable matter by washing with chilled sterile liquor prior to pressing or centrifugation

4. Chilling the yeast and storing it at 0°C prior to transportation

5. Shipping of slurry for short distances (short time span) is acceptable, but a pressed cake of low moisture (around seventy percent) is more suitable for longer distances (time)

6. Smaller quantities can be kept cooler during transportation

7. Monitoring of the entire transfer, from collection of the yeast to use in the recipient brewery, including quality assessments, such as temperature, viability, and bacterial contamination

7

Brewing

The Brewing Process Flow Chart

There are several steps in the brewing process, which include malt milling, mashing, lautering, boiling, cooling, fermentation, maturation, filtration and packaging.

1. **Malt Malting:** To loosen up the endosperm by degrading the endosperm cells' walls and to produce enzymes for further degradation of the content of endosperm cells during mashing.

2. **Milling:** *To provide for optimal conditions for enzymatic activities during mashing for solubilization of fermentable carbohydrates, and to provide for best possible mash separation (and thereby the highest possible yield).*

3. **Mashing:** *To form an extract with a desired profile of sugars, desired level of protein and minor chemical constituents.*

4. **Lautering:** *To separate the sweet wort from spent grain (malt husk).*

5. **Boiling:** *To sterilize wort, deactivate enzymes, concentrate wort, isomerize hop alpha-acid into iso-alpha-acid, reduce volatile compound (such as DMS), increase wort color, reduce wort pH, protein coagulate and produce reducing compounds.*

6. **Cooling:** *To cool down wort temperature to fermentation temperature (from 10°C to 12°C).*

7. **Fermentation:** *To convert fermentable sugar to ethanol, carbon dioxide and beer flavors.*

8. **Maturation:** *To improve/remove unpleasant flavors such as VDK, acetaldehyde, DMS, amyl acetate and so on.*

9. **Filtration:** *To make beer bright (give it a polished shine and brilliance) and more stable.*

10. **Packaging:** *Putting beer into either glass bottles or aluminum cans, after which it will leave the brewery.*

Table 7.1: Sequence of operations encountered in the brewing process

Operations	Input	Product
Milling	Malted/unmalted barley	Grist
Mashing-in	Hot water, cereal adjunct (if any)	Mash
Mash digestion	Enzymes and substrates from grist	Digested mash
Wort separation	Hot water	Sweet wort + spent grains
Wort boiling	Steam, hops, sugar	Hopped wort + hot trub
Wort clarification	Whirlpool tank	Hopped wort + hot trub
Wort cooling and aeration	Refrigeration, sterile air	Cold aerated wort + cold trub
Fermentation	Yeast	Green beer + yeast + cold trub
Lagering or maturation	Refrigeration, enzymes, finings	Carbonated beer ready for filtration
Filtration	Kieselguhr	Bright beer
Packaging	Packages such as glass bottles, aluminum cans	Packaged beer

Brewhouse Procedures

A brewhouse is a building dedicated to the production of hopped wort. It is the most important plant in the brewery. The brewing process divided into seven steps: *Milling (wet or dry), Mashing, Lautering (or mash filter), Wort Boiling, Wort Clarification, Wort Cooling and Wort Aeration.*

The brewhouse is the most important process in beer production. To provide the necessary conditions for this, the initially insoluble components (starch, protein, and so on) in the malt must be converted into soluble products (glucose, maltose, maltotriose, dextrose, amino acid). The formation and dissolving of these compounds is the purpose of wort production.

Steps in Brewing Process:

1. Milling (wet or dry): To create optimal conditions for enzymatic activities during mashing for solubilization of fermentable carbohydrates and to provide for best possible mash separation thus highest possible yield.

2. Mashing: To convert raw materials (unmalted barley, malted barley, and adjunct), in the presence of enzymes, into a fermentable extract suitable for yeast growth and beer production.

3. Lautering (or mash filter): To separate the extracts gained during mashing from the spent grain to clear wort. It is achieved in either a Lauter Tun (a wide vessel with a false bottom fitted with lautering blades) or a mash filter.

4. Wort Boiling: To achieve the following:

   - Flavor development
   - Trub (break) formation
   - Stabilization

- Concentration
- Extraction/transformation of hop components
- Formation/precipitation of protein-polyphenol complexes
- Evaporation of water and other volatiles
- Sterilization
- Destruction of enzyme activity
- Increase in wort color
- Decrease in pH of the wort
- Formation of reducing substances
- Effects on DMS and its precursor SMM

5. Wort Clarification: To separate the coagulated protein (trub) and hop debris from the hot wort.

6. Wort Cooling: To cool down the hot wort to the starting fermentation temperature.

7. Wort Aeration: To aerate (or oxygenate) the wort required for yeast metabolism.

Malt Milling

Objectives of Malt Milling:

1. To break the grain kernels so as to expose the starch endosperm to water, thereby allowing the degradation of the starch by enzymes during the mashing process

2. To cause minimal damage to the husk, and thereby help the wort separation process

Malt milling will be obtained by:

- Crushing the malt husks gently—preferably longitudinally—to expose the endosperm
- Obtaining a complete disintegration of the endosperm to make all of its constituents accessible to enzymatic processes
- Keeping the quantity of fine flour to a minimum to prevent slow lautering
- Preventing extraction of unwanted substances during mashing

Ideal milling is hardly obtainable, but a suitable compromise should be obtained

Types of Malt Milling

1. Dry Milling

Dry milling is commonly performed by roller mills, disk or hammer mills. If the wort separation involves using Lauter Tun, roller mills are employed. Mash filters and vacuum drum filters require a finer particle size distribution and will therefore use a disk, hammer or a modified roll mill. Operation of roller mills requires skilled and trained operators, and frequent quality control for the milled product. All mills are equipped with a feed device, usually a feed roller, to give controlled flow and even distribution of grain into and over the whole length of the milling rollers.

2. Wet Milling

The objective of wet milling is to reduce the mechanical impact on the husks during milling in order to:

- Shorten the lautering process and thereby increase capacity
- Increase extract through increased brewhouse efficiency
- Increase brew length through higher Lauter Tun grain beds as the husk remains intact
- Reduce space requirement
- Retain normal brewhouse operations with poorer modified malts

Problems related to wet milling are:

- Requirement of high-capacity units as the milling of malt must coincide with the mash in operation
- Increased oxygen pick up because of pumping of the milled mash
- High maintenance costs

Wet mills are two-rolls units. The malt is steeped for a defined period of time (ten to thirty minutes) at a preset temperature (27–49°C) prior to being milled. The mash in program is a function of the milling duration. After completion of the milling operations, the remaining mash water is run through the mill to flush the system clean.

Milling of malt depends on:

- Modification—under-modified malt gives coarse grist
- Moisture content—low moisture malt gives finer grist (normally less than four percent)
- Lautering equipment—for Lauter Tun coarse milling and for mash filter semi coarse milling
- Final brewhouse yield—coarse milling will give high brewhouse yield with Lauter Tun
- Lautering time (for Lauter Tun only)—fine milling leads to longer lautering time

To optimize the brewhouse process, it is critical to obtain the greatest degree of extract from the mash while ensuring the minimal amount of time for the mash separation in the Lauter Tun or mash filter. Minimal destruction of the husk (for Lauter Tun) during the milling process is important for both wort and beer quality.

The finer the grinding of the husk, the greater the tendency to extract fatty acids, phenolic compounds, and glucans. These can have a negative influence on the beer both chemically and organoleptically.

Practical Aspects of the Milling System

Quality control of the milling system is critical from the standpoint of safety, good manufacturing practice, and quality of the product. Quality of the grist should be verified by the efficiency of the brewhouse operations, productivity (number of brews per day), and as a double check through analysis of residual grains. Analysis of the grist composition should be scheduled on a weekly basis.

Mills should be cleaned at the end of the production period—the magnet can be cleaned off at the same time. Inspection should be made for any potential insect activity. Fumigation or fogging of the mill is best done on regular schedule. Gap and parallel position of the rolls should be checked weekly.

Mashing

Mashing is the process in which malt or malt and adjunct are mixed with water in a controlled heating process to digest and extract proteins, carbohydrates, enzymes, and phenolic substances to obtain fermentable sugars and nitrogen compounds for yeast nutrients. By controlling the mashing process throughout, the brewer can achieve a proper balance of carbohydrates and proteins that are necessary for producing beer and avoiding other species of carbohydrates and proteins that may cause problems in the process or beer. Any errors made in the mashing processes are not easily corrected and can render the following brewing process very difficult. Therefore, brewers should pay close attention to these operations.

The Aim of Mashing

The aim of mashing is to form an extract with a desired profile of sugars, proteins and other minor chemical constituents. Most of the extract is produced during mashing by the action of enzymes, which are allowed to act at their optimum temperatures.

Raw materials quality may vary with the season, and availability can be unstable, but the beer has to be kept flowing nonetheless. Brewers all over the world are pushing the limits for mash separation and beer filtration, because these two operations are critical when it comes to maintaining a high and constant production flow in spite of fluctuations in the quality and availability of raw materials. Many brewers also wish to supplement the malt with barley.

A number of enzymes for brewing help brewers in cutting down production time and cost while delivering the quality product that consumers have come to expect. Such enzymes need to perform highly specific tasks as well as, or even better than, naturally occurring enzymes.

The brewing enzyme actually consists of two enzymes:

- β-gluconase
- Xylanase

This will help brewers become less vulnerable to fluctuations in raw materials quality and availability. Also, the mash separation process typically constitutes a bottleneck in the production flow, and an enzyme that can reduce viscosity in the wort will help overcome this problem. The β-glucan fraction in the barley and malted barley, which gives a headache at mash separation and beer filtration, is the β-1,3-1,4 glucan.

In brewing, β-glucan is solubilized during the entire mashing process all the way to mashing-off. The enzymes used to break down the β-glucan—the β-glucanases—need to have a thermostability profile that fits the mashing profile.

Mashing is the most critical operation in the brewhouse. Within the constraints of the malt and adjunct quality, the mashing process will directly determine:

- Alcohol content of the beer
- Concentration of the unfermented sugars in the beer
- Peptide and amino acid profile of the beer
- Yeast nutrients for vigorous fermentation
- Buffering capacity and pH of the wort and beer
- β-glucan content of the beer
- Efficiency of the extraction of sugars

In addition to the above, the mashing operation influences the physical properties of the beer, such as foam, color, and clarity.

Water used for mashing should be free of bicarbonate hardness (≤50 mg/l). Calcium salts are often added to the mashing water in the form of calcium sulfate (gypsum) or calcium chloride to achieve a level of 100–150 mg/l of calcium. Calcium is a coenzyme for α-amylase, and controls the pH of the mash and the calcium ion in the beer during lagering (25–75 mg/l). It also precipitates oxalate as the calcium salt. Oxalate is the undesirable y-product of fermentation and must be removed. Brewing water is used at the rate of 3–3.7 liters per kilogram of the grist. Thin mashes favor hydrolytic reactions such as hydrolysis of carbohydrates to fermentable sugars, while thick mashes solubilize more nitrogen compounds.

Mashing Process in Brewing

Protein Rest

The initial heating of mash is about 43–49°C for a period of ten to forty minutes. This is commonly called the *protein rest* and, during this time, the malt and adjunct become hydrated. In well-modified malts, the simple peptides and sugars are solubilized and the malt proteases hydrolyze the protein to simple peptides and amino acids. The extent of this proteolysis depends upon the temperature (lower temperature favor hydrolysis) and time (longer time increases hydrolysis).

During protein rest, the β-glucans from the starch cell wall matrix of the malt are hydrolyzed by the β-glucanase, which is also derived from the malt. These enzymes have an optimum temperature of about 40°C and are inactivated at 60°C. Beta-glucans increase wort viscosity, which increases run-off from the Lauter Tun and also can cause problem in beer filtration later in the process. The extent of the problems from β-glucans can be diminished by the extent of malt modifications.

Conversion Rest Temperature

Protein rest is followed by a rise in temperature to the conversion rest temperature (low conversion temperature of 65°C at the rate of 0.5°C rise/min, and high conversion temperature of 71°C at the rate of 1.5°C rise/min), which is also called 'the ramp'. During this phase, the hydrolysis of starch to dextrin and to fermentable sugars is initiated. These reactions are brought about by two enzymes:

1. Alpha-Amylase, which solubilizes the starch and hydrolyzes the amylase to limit dextrins and some glucose. In addition, it also hydrolyzes amylopectin to maltose.
2. Beta-Amylase, which hydrolyzes amylase to maltose and amylopectin to limit dextrin.

These enzymes have different temperature optima and inactivation. Alpha-Amylase has an optimum temperature at about 70°C whereas it is inactivated at about 74°C. Beta-Amylase optimum is at about 63°C and inactivation occurs at about 68°C. Thus during the rise to conversion temperature, six major reactions occur:

1. Hydrolysis of glucans
2. Denaturation of protein
3. Activation of α- and β-amylase
4. Inactivation of proteinases
5. Inactivation of glucanases
6. Inactivation of β-amylase

Once the conversion temperature is reached, the mash is generally held for thirty to forty minutes. This 'conversion rest' generates the maximum yield of extract from the malt to be solubilized and conditions the mash for extraction in the Lauter Tun. To verify whether all the starch has been hydrolyzed, this is commonly done by means of an 'iodine test'. Unhydrolyzed starch will cause haze in the final beer.

Mash Off:

This is the final phase of the mashing operation, in which the temperature is quickly raised to 77°C at the rate of 1–2°C/min to inactivate the carbohydrate enzymes and fix the amount of fermentable sugars.

Mash Tun/Kettle Operation:

Quality mashing operations result in good brewhouse efficiencies and favorable final product characteristics. The biochemical reactions that take place in the mash kettle help in determining the product flavor, body, shelf life and innumerable other characteristics necessary for a good, consistent product. The agitator operates during mash in, application of heat, and mash transfer. It does not operate during rest periods. Water correction minerals in the form of calcium sulfate (gypsum), calcium chloride or acids are added prior to the grist in the foundation mash water.

Wort Separation/Lautering

Wort separation is the method of separating grain particles from liquid extract (wort) without altering the chemical make-up of the mash. In practice, it is not only separation of wort from the grain but also the extraction of solids from the spent grains. The extraction efficiency is measured as 'extract yield', which is the ratio of the mass of the extract to the mass of the malt or malt and adjunct.

This operation is performed either in a Lauter Tun or a mash filter. The selection of either device is based on economics, process configuration, and available space in the brewhouse. In the *Lauter Tun*, the malt husk forms a filter bed to support the mash. This requires that the milling operation crush the malt rather than grind it to keep the malt husk intact. In the *mash filter*, the filter cloth or membrane provides the media or septum to support the mash; hence, malt may be milled finer for better extraction.

Lauter Tun

The Lauter Tun is internationally accepted and continues to be the predominant wort separation device for brewhouse operations. A brewing day is normally taken to be based on continuous 24 hours operation, and the Lauter Tun or other filtration device is generally rate-limiting step, defining the output of the whole brewhouse.

Objectives of Lautering

The main objectives of a Lauter Tun operation are:

1. High level of extraction in an efficient manner
2. Clear wort with low solids

3. Extract/repeatable cycle time

4. Wort with low dissolved oxygen

5. High mechanical efficiency

6. Low moisture spent grains

7. Minimum effluent disposal

The Lauter Tun is round, a flat-, valley-, or slightly sloped-bottomed vessel. The vessel has a vent for relieving the vapors of the hot mash to the atmosphere. The filtering system itself is a milled slotted steel false bottom. There are run-off ports for collection of the lautered wort. To assist mash transfer and distribution in the Lauter Tun, arms with vertical knives are strategically placed for leveling the bed and facilitating filtration of the liquid from the mash. The mash in the Lauter Tun must be washed during the lautering process to recover all the extract desired. This washing is known as *sparging*. A sparging system is installed in the Lauter Tun, spays high temperature (77°C) water over the mash. This dilutes the sugar solution and washes out the extract. The filtered wort is collected via a series of lines and ultimately goes directly to the wort kettle. When the lautering operation is complete, the remaining water is drained and the spent grain is discharged to a dump tank, where it is sold (in a wet state) as cattle feed.

Basic Components of a Lauter Tun

1. Vapor stack for release of water vapor generated during mash transfer, lautering, sparging, and grain-out process

2. Lauter Tun dome, which retains the vapors and prevents heat loss; the domes are often insulated to further reduce heat loss

3. Lauter Tun real bottom, where the filtered wort is collected; real bottoms are either flat or with valleys

Lautering run-off ports and nozzles are mounted in the real bottom to flush and spray it clean with hot process water between cycles. The characteristics of a Lauter Tun are:

1. Wort separation is completed through a milled slotted bottom. Free surface area (the area through which wort can flow) varies from eight to twelve percent of the total surface area. Loading on the false bottom is critical for long shelf life of the Lauter Tun. Dry milling ranges from 163.6–279.8 kgs/m²; conditioned dry milling 205.1 kgs/m²

2. The sides, and often the domes, of the Lauter Tun are insulated to retain heat. Heat retention maintains low viscosities and inhibits microbial growth

3. Mash transfer lines from the mash kettle are directed either through the side or bottom of the Lauter Tun reduces turbulence and oxygen uptake

4. The lautering machine serves:

    a. To level the grain bed during mash transfer, if required

    b. Loosen the grain bed during run-off

    c. Facilitate spent grain discharge at the end of the lautering process

5. Lauter Tun drive is either hydraulic or mechanical. Mechanical drives have three speeds:

    a. Slow   : 1/3 rpm for lautering

    b. Medium   : 3 rpm for mash transfer

    c. High   : 9 rpm for spent grain discharge (reverse direction)

6. A hydraulic-lift cylinder lifts and lowers the lautering machine during the lautering process and throws the spent grain out. The operated is based on clarity and differential pressure

7. The sparge water system is a series of circular pipes suspended from the Lauter Tun dome. Nozzles are installed in the pipes to promote a uniform spray of hot water at 77°C across the surface of the Lauter Tun bed. There is a provision for adding foundation water and underlet water at 77°C at the initial phase of the overall lautering process

8. Lautering channels facilitate wort collection—these lines are routed singly or collectively to concentric collection rings under the real bottom

9. Wort flow rate is controlled at the collection point manually or through process controllers. Modern Lauter Tuns have turbidity and density meters installed on line

10. The flow control valve is adjusted for each step of the process, the major steps are:

    a. Re-circulation or forerun

    b. Run-off

    c. Drain out

11. The spent grains discharge door allows the timely discharge of spent grains. A Lauter Tun has a spent grain-out door that is manually, mechanically or hydraulically operated. For optimum production, it is critical that minimal time be required on spent grain discharge

12. The dump tank for spent grain discharged from the Lauter Tun permit rapid spent grain-out. The dump tank is designed to hold 1.5 to 2 brews

Cycle Time

The cycle time of a Lauter Tun's turnaround time for a rapid batch brewhouse operation is calculated by dividing 24 hours/day by the desired number of brews/day.

    The cycle time must accommodate each of the following steps, against which specific time should be allowed.

- Plate flood (*)
- Mash transfer (*)
- Wort re-circulation
- Wort collection
- Bed drain-down (*)
- Spent grains discharge (*)
- Under-plate flush/rinse tun (*)

The steps marked (*) do not contribute directly to the process and are classified as non-productive, and are therefore to be minimized without compromising the process. Wort re-circulation has an impact on wort quality, extract efficiency and cycle time, since this element is a productive phase of the cycle.

Mash Filter

Although Lauter Tuns are widely employed for wort separation, some large volume breweries prefer mash filters. These are very much like plate and frame filters, consisting of a series of grid-type plates alternated with hollow frame plates that are suspended on side rails. Mash filters are more likely to be used in larger breweries where throughput and floor space are priorities.

Mash filters perform the wort separation in a thin filtration bed process. Filtration is initiated with the transfer of the mash to the plate and frame system. Modern mash filters have filter mediums made generally of polyethylene or polypropylene fiber, and are selected based on quality of filtrate. The unit is hydraulically closed so that it is fluid-tight.

The mash filter's capacity is determined by the depth of the bed, length and height of the filter frame; the number of the frames are limiting capacity factors for both the mash filter and size of the brew.

In the mash filtration system, mash from the mash kettle enters the plate and frame assembly from the top. Sparge water is pumped into the unit from the bottom. Filtered wort exits the filter from one side. To initiate the process of mash filtration, the mash filter is flushed and preheated with hot water at 77°C. The mash is then pumped into the filter, and the entire mash transfer to the mash filter is completed in approximately twenty to thirty minutes. As soon as the filter chambers are hundred percent full, the wort outlet valves are closed. If the chambers are not full, the efficiency of the extraction process suffers significantly because the sparge water will flow through the empty portion of the chambers. When all the mash is in, the wort collection system is opened, and the first wort is drawn. The wort is re-circulated through the filter until desired/satisfactory clarity is achieved.

Sparge water is pumped into the filter from the bottom and to every other wort collection chamber. This causes the water to be pumped through the grain bed and out of the opposite wort collection chamber. The flow rate of the sparge water is matched with the flow rate of wort and increase in differential pressure. When is wort kettle is full, the filtering process is ended.

The filter is then opened and the plates are separated for the spent grains to drop from the frames into the grain collecting system.

Wort quality remains comparable to that of the Lauter Tun with respect to total solids; however, the cycle time on wort separation is significantly reduced.

In classic mash filters, the cotton cloth had to be cleaned after every cycle; the polypropylene cloths used today require high pressure cleaning only once a week. Mash filters can be filled with one to two percent sodium hydroxide solution and circulated at the end of each week.

Technology of the Meura Mash Filter

Standard process steps of the Meura Mash Filter are:

1. Filter fill cycle
   - The filter is filled from the bottom inlet valve under low pressure
2. Filtration
   - The filtering operation develops a self-filtering grain bed on the filter cloth
3. Pre-compression Phase
   - The chambers are expanded with air pressure and cause a low pressure on the grain; first wort is thereby collected prior to sparge water
4. Sparging
   - Sparge water is pumped into the grain chamber through the mash inlet line. Distribution is uniform across the entire filter bed
5. Compression
   - Compression is once again done with the air, which forces out the second wort
   - Spent grains are discharged with a fully automatic device

Strainmaster

The Strainmaster, which is not as widely used as the Lauter Tun or the mash filter, was patented by Anheuser Busch and developed by the Nooter Corporation.

The Strainmaster consists of a rectangular hopper-bottomed tank that is fitted at the base of the hopper and that has downward opening doors for the discharge of spent grains. Within the vessel are header pipes running longitudinally down its length, and attached to these are perforated straining tubes. The header pipes are connected by wort draw-off mains to wort pumps. Wort withdrawal and circulation is initiated when the top row of tubes is covered. The wort is then recycled through the straining tubes to the top of the Strainmaster. The action of drawing wort through the tubes creates a filter bed around the tubes in much the same manner that occurs relative to the Lauter Tun. Re-circulation continues until the desired wort brightness is achieved, and then the wort is collected in the wort kettle.

Comparison of Separation Systems

A comparison of the Lauter Tun, mash filter and Strainmaster showed that the process costs for the Lauter Tun and Strainmaster declined with increasing capacity, whereas it increased in proportion to its capacity for the mash filter. Maintenance costs are considerably lower for the Strainmaster than for the mash filter and the Lauter Tun.

Wort Boiling

Wort boiling is the key operation in the brewhouse, where a number of brewing or process functions that are essential to produce a quality finished product are carried out through the downstream processes of fermentation, lagering/maturation and so on.

We should consider the design and engineering of the wort kettle (copper) in its modern form and their benefits from the viewpoint of product quality. It is always a vertical cylindrical vessel, usually with a dished bottom, typically with a height (wort depth) to diameter ratio of around 1:1.

Wort Boiling System

Brewers make concerted efforts to minimize the impact of the brewing process on the environment. For energy conservation, wort kettle boil times are being reduced from over two hours to one hour. Vapor recompression systems or vapor condensers are installed to minimize evaporation to the atmosphere. To maximize the productivity of a brewhouse, vessel cycle times are held to a minimum. An effective CIP system along with the schedule is programmed to clean the wort kettle at regular intervals.

Steam heating systems have been proven to be the most efficient, economical and valuable method of wort boiling. There are five types:

1. Internal boiling systems
2. External boiling systems
3. Low-pressure boiling
4. High temperature boiling
5. Continuous high temperature wort boiling systems

Briggs has designed the *thermosyphon wort boiling system*, which is a modern external wort boiling system consisting of tube and shell heat exchanger called 'kalandria', where thermosyphon or natural circulation flow through the axial flow and by-pass the circulation pump.

Advantages:

- Ability to start heating and boil at low volumes
- Effective CIP
- Generated better hop utilization
- Saving in electrical energy usage and pump wear and tear
- Vigorous two phase flow
- Nucleate boiling and minimum shear

Objectives

The main objectives of the wort boiling process are:

1. Sterilization of the Wort

Brewing raw materials such as malt, hops and occasionally brewing water itself are infected by microorganisms, and these have to be killed during the brewing process to prevent wort and beer spoilage.

After boiling, the wort is largely free from microbial contamination. Some microorganisms, primarily the *Bacillus Species* and other thermophilic bacteria, are able to form spores that can withstand heat treatment, including boiling, and, if present in raw materials or the brewing water, may persist in the finished beer. However, beer does not support the subsequent growth of these organisms.

2. Halting Enzyme Action

Enzymes rely on their three-dimensional structure for their activity. Above certain temperatures, usually in the range of 50–70°C, the tertiary structure of the enzymes becomes denatured, and they lose their activity. By the time the wort has reached boiling point, there is usually no residual enzyme activity.

3. Concentration of Wort

During wort boiling, water is driven off as steam, thus concentrating the wort. The amount of water removed during the boil is directly proportional to the rate of evaporation once boiling has been achieved. The efficiency rate is affected by the design of the wort kettle, particularly the surface area.

4. Isomerization of Bitter Hop Compounds

During boiling, the insoluble alpha acids extracted from hops are converted to a more soluble iso-alpha-acid. This reaction is accelerated by temperature.

Isomerization is a relatively rapid reaction with production of over ninety percent of the wort bitterness occurring within the first thirty minutes of boil. Maximum isomerization usually occurs within sixty to seventy minutes of boiling and accounts for around sixty percent of the total alpha-acid present. Iso-alpha-acid continues to be lost during the fermentation and maturation process and is lost in any foam produced so that final conversion value of alpha-acid into iso-alpha-acid in the beer is around forty percent.

5. Removal of Volatiles

During the evaporation stage of wort boiling, undesirable volatile compounds (DMS or dimethyl sulfide—the taste of which is described as 'sweetcorn', hop oils and so on) are driven off with the steam.

Principal factors that affect the evaporation of volatiles include:

- Temperature of wort
- Vigor of boil
  - o Surface tension
  - o Condensation of volatiles in the vapor stack
  - o Thickness of diffusion path
  - o Duration of boil

The design of the wort kettle will have a major influence on the factors listed above, and it has been found that more late hop character persists in gently agitated system than in more vigorous boiling systems, such as isometric kettle, than in more vigorous boiling systems with turbulent flow, such as wort kettle fitted with external wort boiler.

6. Increase in Color

The color of the wort increases during the boil. The reactions responsible for color development fall into three broad categories:

- Maillard reaction between carbonyl and amino compounds
- Caramelization of sugars, which is limited in steam-heated wort kettles
- Oxidation of polyphenols

Oxidation during wort boiling increases the color, particularly with the oxidation of the polyphenols, which also has the effect of decreasing the reducing power of the wort and beer.

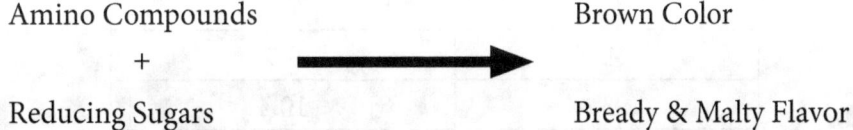

Mash and wort produced with low oxidation produces lower wort and beer with lower colors and improved flavor stability.

7. Reducing Wort pH

Control of pH throughout the brewing process, from brewing water to the final package, is fundamental for finished product consistency. Wort pH starts to decrease during mashing and continues to fall during wort boiling. The principal fall in pH is due to the reaction of Ca_2^+ compounds with phosphates and polypeptides to form an insoluble compounds releasing H^+.

At least half of the calcium present in the wort is precipitated by the end of boiling. Hence, sweet wort with a starting concentration of 100 ppm (100 mg/l) will produce beer with around 40 ppm of calcium.

To assist in the fall in pH, extra calcium ions in the form of calcium sulfate or calcium chloride are added to the wort kettle. An alternate method to decrease pH is through the direct addition of acids such as phosphoric or lactic acid to drop the wort pH.

It is important to achieve the required decrease in pH (around pH 5) as it affects wort and beer character. In particular, the fall:

- Improves protein coagulation
- Improves beer flavor in particular VDK reduction
- Encourages yeast growth
- Inhibits the growth of many other contaminating organisms
- Lower pH results in poorer hop utilization
- Lower pH results in less color formation

8. Reducing Wort Nitrogen Levels

During the brewing process, it is necessary to decrease the level of high molecular weight nitrogen, which comes from the malt. If allowed to persist, it can affect the pH, colloidal stability (chill haze and permanent haze), fining and clarifying properties, and fermentation and taste of the beer. Wort boiling is only one, important, stage, in the reduction of nitrogen, and the effects of reducing the amount of wort nitrogen for a standard boil at 100°C are shown below:

Time of Boil (hrs)	% Nitrogen Removal
0	0
0.5	5.4
1	6.2
1.5	7.7
2	9.9
3	10.4

Because of the relatively small overall reduction in total nitrogen during boiling, it is difficult to obtain consistent results even from the same wort kettle with the same quality of wort.

The degree of protein and polypeptide removed depends on the probability of the individual molecule colliding and forming stable bonds during the boil, and this is directly proportional to the length and vigor of the boil for a given temperature.

Traditional criteria used for evaluating efficient wort boiling are:

- Temperature of boil (usually just above 100˚C when boiling under atmospheric pressure)
- Length of boil (ninety minutes)
- Evaporation percentage per hour (ten percent)

However, because of the need to reduce energy costs and to improve brewhouse efficiencies shorter boiling times with lower evaporation rates are now employed.

Typical modern wort kettles operate with sixty-minute boil, between five to nine percent.

With adequate turbulence during boil, the actual removal of the high molecular weight nitrogen fraction is a function of time and vigor, and can be relatively independent of evaporation rate for atmospheric boiling.

Vigor is only one feature of importance for coagulation, since protein agglomeration is improved by intense vapor bubble formation. The actual wort surface temperature, and the duration of the intimate contact of the wort with the heating surface, may also be of importance.

It is desirable to remove as much protein/polypeptide as possible; nitrogen compounds have an important role in the quality and fermentation performance of a beer and in providing foam compounds and mouth-feel. Excess protein/polypeptide removal could lead to a finished product of poorer quality.

9. Extraction and Precipitation of Tannins/Polyphenols

Simple hop tannins and most malt polyphenols are soluble in boiling wort and moderately soluble in cold water. Tannin/polyphenols also combine with proteins to form protein/polyphenol complexes:

- Proteins that combine with oxidized polyphenols are insoluble in boiling wort, and are therefore precipitated during the boil to hot break
- Proteins that combine with unoxidized polyphenols are soluble in boiling wort but precipitate when chilled and can give rise to chill haze and cold break. The polyphenols may subsequently oxidize during beer processing and may produce colloidal instability in packaged beer

Unprocessed hops contribute around forty percent of the total polyphenol content to boiled wort; however, most hop polyphenols are removed due to hot and cold break. The rest of the polyphenols comes from the dry goods (particularly the husk), and are less polymerized—hence, are less likely to be removed. Worts lack of hop tannins give poorer wort clarity and have a lower reducing potential.

10. <u>Producing Reducing Compounds</u>

A number of reducing compounds that come from the raw materials, such as tannins, reductones and melanoids, are formed during wort boiling through the condensation between sugar and amino compounds. Addition of high amount of unprocessed hops will tend to produce high reducing power compounds. Brewing systems with low levels of oxidation tend to preserve the natural reducing compounds in the worts, which can persist into packaged beer and delay the onset of maturation, thus improving colloidal and flavor stability.

The New Meura Concept with Wort Stripping

Increasing energy prices made breweries think continuously about energy saving. During wort production, about thirty-five percent of the energy in brewery is spent. This is exactly where the brewery began to take matters into its own hands. A new concept for wort boiling and clarification was introduced in order to save energy and to improve the quality of the end product. The concept is to optimize the trub and to eliminate volatile substances more efficiently. The plant consists of a formation tank or wort kettle, a hot break sedimentation tank or whirlpool, a stripping column and wort cooler.

The wort is kept in the wort kettle at 100°C for approximately thirty to fifty minutes without any noteworthy evaporation. Afterwards, it is pumped into the sedimentation tank or a whirlpool. After the settling time or whirlpool rest the stripping starts. The inside of the stripping column consists of approximately 5 cm large stainless steel rings in order to obtain a surface that is as big as possible. The wort flows from top to down. Vapor is injected in the reverse flow, and thus evaporation of DMS in particular is guaranteed. The stripping is directly followed by wort cooling so that no further free DMS can emerge.

The result of the operation is very positive. Less energy is necessary to eliminate unwanted aromas like DMS. With an evaporation of only 2.5 percent and 0.5 percent stripping vapor, a more efficient evaporation of free DMS can be achieved. Amongst others, aldehydes could be better removed than with ordinary wort boiling. Rise of foam stability and color increase during boiling < 1 percent EBC could be noticed.

There are several benefits to the New Meura System:

1. Reduction of the primary energy at wort boiling up to fifty percent, because it is possible to reduce DMS (dimethyl sulfide) additionally to the minimized boiling process
2. Reaction regarding bad malt quality, resulting in higher DMS content after the wort boiling, keeping the boiling system on a constant low energy level followed by wort stripping
3. Reduction of the boiling time accessory to the minimized energy boiling process

A new concept featuring a reduced carbon-footprint, easy retrofit options and very flexible system can be integrated very easy in existing boiling systems, between the whirlpool and wort cooling system.

Wort Kettle Additives/Addition to Boiling Wort

Consistent desired quality and good clarity of wort has always been regarded as an essential quality from the brewer's viewpoint. Clarifiers have been traditionally used in the brewhouse to clarify wort.

With reference to wort, a clean and clear fermentable substrate has always been used by brewers to achieve a consistent, high quality product. A desirable characteristic of hot break formation during wort boiling is to have rapid flocculation with strong cohesion and compactness of the trub in the whirlpool, allowing clear wort production.

The mechanical action of the boiling wort in the wort kettle presents opportunities for the brewer to add the followings materials into the process for numerous purposes:

Hops

Timing

The hops may be added at the start of the boil or near the end of the boil. Bittering hops are usually added at the beginning of the boil to maximize isomerization of their alpha-acid content and to drive off undesirable flavor compounds. Aroma hops are added part way through or at the end of the boil for the delicate hop flavor components. Some bittering hops contribute to hop aroma, even when added early in the boil.

Finish Hopping

The primary method used to get hop flavor and aroma in the beer is to add hops very late in the boil. This practice is called 'late hopping', 'late kettle-hopping', 'aromatic hopping', or 'finish hopping'.

Dosage Procedures

In small breweries, whole hops or pellets are manually added to the kettle. When hops are added to the kettle, the heat should be shut off or the hops added very slowly to prevent a boil-over.

Rates

The quantity of hops is determined by several factors—alpha-acid content, desired hop flavor and aroma in the finished beer, condition of the hops, efficiency of the hop extraction, brewing process, and type of brewing water. The hopping rate for bitterness can be calculated, whereas the rates at which hops should be added for flavor and aroma are less accurate.

Kettle/Copper Finings

One way of enhancing floc formation is through the addition of kettle finings. Copper finings (also referred to as 'Irish moss') is a form of seaweed consisting mostly of a complex starchy polymer called K-carrageenan. This, like polyphenols, has a negative charge and is effective in precipitating positively charged proteins from the wort solution.

Dosage Rate

The rate of use of copper finings varies widely from brewery to brewery but is generally found in the range of 10 to 80 mg/l (or ppm). The major reasons for the widely differing rates of use are variations in grist formulations, mashing systems, and gravities of the wort.

Choice of Material

Kettle or copper fining products come in a range of physical forms and degrees of purification, including powders, tablets, granules, refined carrageenan, and alkali-washed seaweed.

Timing

Timing the addition of copper finings is essential, because K-carrageenan denatures and becomes less active with prolonged treatment at high temperature.

Dosage Procedures

Powdered products are usually thoroughly mixed in cold water to provide a suspension that can be added directly to the kettle or the whirlpool. If powdered products are added directly to the kettle, there is likelihood that a significant amount of product will be sucked up the chimney, which results in under-dosing and loss of performance.

Caramel

Some brewers add caramel to achieve the desired wort color, and, hence, targeted finished beer color.

Lactic Acid

Lactic acid can be added to lower the pH of the wort.

Calcium Sulfate or Calcium Chloride

Calcium sulfate (gypsum) or calcium chloride is often added to lower the pH of the wort during the boil.

Syrups and Sugars

Sugars can be added to the kettle in either dry or liquid forms.

Hot Wort Separation/Clarification

The objective of hot wort separation is to receive the wort from the wort boiling stage and, by the application of sedimentation, extract the insoluble elements (usually hops/trub) to allow bright wort with low solids to be passed to the next stage of wort cooling. The objective should be achieved with highest efficiency to ensure optimum wort quality.

It is very important to remove spent hop materials and trub since they serve no further purpose or value. These materials can be detrimental to beer quality, processing efficiency and the cycle time of the downstream stages, in the following ways:

- Cause blockages at the wort cooler
- Hinder wort clarification
- Trub coats the yeast cell and hinders fermentation
- Lipid from the trub will be detrimental to beer head retention
- Trub increases the quantity of break-rich materials and losses at fermentation end
- Trub makes downstream beer filtration more difficult

On the completion of wort boiling, it is necessary to separate the hop debris and coagulated protein (hot break) from the boiled wort before cooling. The hot break or trub is made up of:

- Coagulated protein or 'break', which, if allowed to remain, would cause flavor and haze problems in the finished beer
- Tannin (polyphenols) material from the malt husk and from the hops, which combines with proteins to form chill haze and then permanent haze in the finished beer. In higher concentration, it can cause astringency
- Lipids or fatty material from the hops and wort that will destroy the beer's foam (head stability) and will contribute to beer staling leading to the early formation of stale papery/cardboard notes
- Hop debris that may be in the form of spent whole hops or hop pellets

Insufficient trub removal or defective break formation can cause the passage of a large proportion of the sludge to the fermentation system. Defects that may result include poor yeast performance with an unsatisfactory attenuation. The fermentation may be slow and yield high final gravity, the beer may be open to infection and slow to fine, a persistent haze and sulfury notes in the final beer may result.

The recognized modern systems to be considered and examined are:

- The whirlpool
- The hopback
- The hop separator
- The hot wort centrifuge

The whirlpool

The whirlpool tank is a predominant device in modern installations for removal of the hot break. Use of hop pellets, powder or extract at the boil is mandatory since the whirlpool cannot handle whole leaf hops.

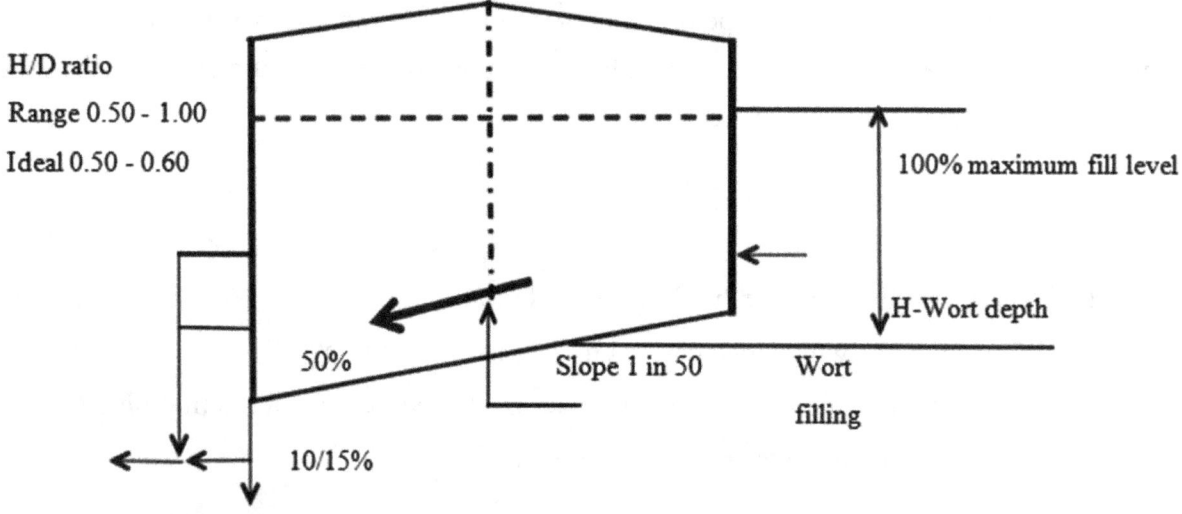

Figure 7.1: Flat Bottom Sloped Floor Whirlpool Design

The whirlpool is an enclosed circular vessel, usually fabricated from stainless steel and constructed to a defined height to diameter (H/D) relationship to the wort surface. There is a growing trend towards the defined range of 0.5–0.6 for new installations. This trend is understandable since it is appreciated that the wort is cast into the vessel almost tangentially; therefore, angular rotation of the wort occurs and centrifugal sedimentation forces come into play.

Charging of the vessel at the tangential/inset nozzle can impart shear force into the wort and destroy the flocculation that has started once the heating/evaporation has been stopped in the kettle. It should be appreciated that the manner in which the wort is pumped into the whirlpool is particularly important, in order to minimize shear force and promote optimum conditions to settle and capture the trub. The pump-in time and wort velocities within the system should be critically examined for optimum trub settling performance.

The advantage of using the combined kettle/whirlpool design option should be considered as a particular benefit in avoiding trub disruption and wort shear, since the casting stage is eliminated.

Whirlpool designs and configurations fall into two basic categories:

- Cone-down or sump bottom to capture the trub, from where the trub may be continuously separated as it forms at the whirlpool stage using an external centrifuge
- Semi-flat bottom, which is usually sloped one in fifty for drainage, which allow trub cone draining and disposal of the debris with water

Of the various shapes used, the conical and flat bottom designs appear to be the preferred designs.

The hopback

The hopback is the traditional hot wort clarification device from the era when whole leaf hops were dominant. Due to the decline in the use of whole hops, the inclusion of this technology into new schemes is becoming a rarity. Use of the whole leaf hop is mandatory with this device to allow the hot break to be removed by filtration in a single stage operation.

The traditional hopback is a circular vessel fitted with the following equipment:

- Wort inlet distribution box
- Water driven rotating tangent sparger
- Under-plate void venting pipe
- False bottom system
- Rotating non-rinse spent hops discharging blade
- Wort collection direct to PHE
- Spent hops discharge valve
- CIP devices
- Man door with viewing glass

The hopback is sized to contain the total cast wort volume from the kettle. The spent whole leaf hops materials are in fluid suspension with the wort when it is cast into the hopback with the wort, together with the trub (hot break). The hop leaves quickly settle at the bottom of the vessel to form the filter bed.

The whole leaf filter bed, which is in loose fluid suspension with the wort, traps the hot break materials in the interstitial voids and, as such, must not be cast down to the wort cooling stage at too fast a rate or with any substantial amount of differential pressure—otherwise, the filter bed properties would be destroyed.

When the filter bed has formed, the wort may be re-circulated until it is fully bright before starting the casting to the wort cooler. At the latter stages of pumping to the wort cooler, and when the hop-bed is revealed, the bed is sparged with hot brewing water using the tangent sparger device to recover the entrained wort/extract left in the filter bed.

After casting and sparging has been completed, the hopback may be left standing, waiting for the next brew, which is cast onto the previous spent hops material. Alternately, the hopback is drained and discharged of the spent hops using the rotating blade device before being rinsed and cleaned.

The hopback is the traditional device for hot wort separation when whole leaf hops are used. The hopback offers considerable flexibility in handling recipes with different whole leaf hop addition rates. The traditional element and the decline in the use of whole leaf hops do, however, limit the number of suppliers with the necessary design expertise and capability to design the system.

The hop separator

The hop separator was a development that occurred alongside that of the whirlpool, especially to remove the whole leaf hops that the whirlpool could not handle. As the name implies, the machine only separates the whole leaf spent hop materials from the hot wort as it is cast from the wort kettle. Separation of the hot break materials cannot be accomplished in this device. The trub removal is usually performed downstream, using either a whirlpool or hot wort centrifuge.

Use of whole leaf hops is mandatory when using the hop separator. Due to the move away from the use of whole leaf in preference to pre-processed hop materials, the hop separator is likewise in decline and rarely used in any new scheme.

The hop separator is a freestanding machine. Its components are:

- Wort inlet control valve and level control system
- Geared motor unit driving an inclined Archimedean screw
- Strainer basket section below the inclined screw
- Compression section of the screw section
- Sparging section of the conveyor
- Compressed air injection and spent hops blower
- Wort discharge to whirlpool or hot wort centrifuge
- Clean-out valve

The hop separator receives wort cast from the wort kettle, which has spent whole leaf hops and trub flocculation in fluid suspension. The incoming wort is continuously received into the hop separator chamber using an automatic level control system.

The bottom of the chamber has a rotating inclined Archimedean screw that is encased by a perforated plate through which the bulk of the wort passes before leaving the machine for the removal of the hot break. The spent leaf hops material is transported to the end of the conveyor, where it is sparged with hot water and compressed to remove the residual extract (which is combined with the other wort) before being ejected as a waste by-product.

The hop separator offers little flexibility and can easily be completely removed from the stage by simply adopting one of the modern processed hops addition products in the wort kettle and pumping directly into the whirlpool.

The hot wort centrifuge

The opening bowl disc centrifuge is exclusively used for removal of the hot break. This type of machine cannot handle whole leaf hops, and the hot wort centrifuge can sometimes be found operating in conjunction with the hop separator for removal of the hot break. However, few breweries are using disc centrifuge for processing the hot wort from the wort kettle where hop pellets or concentrated hop extract has been used.

The hot wort centrifuge is a freestanding machine; its components are:

- Base plate
- Large electric motor
- Gear box
- Belt drive
- Robust bearing assembly
- Wort infeed from the wort kettle
- Hydraulically operated bottom bowl
- Solids ejector when the bottom bowl opens
- Clarified wort out to the underback

Hot wort from the wort kettle is fed into the machine at a controlled rate. The centrifuge bowl contains about 200 conical discs that are stacked vertically and rotated at very high speed (about 600 rpm). The discs contain ascending holes/slots at about mid-radius to allow the incoming wort to be distributed evenly through the stack. The clarified wort migrates inwards and the solids migrate outwards to be ejected as sludge. The clarified wort leaves the machine to be passed to a holding vessel before the wort cooling stage.

In reality, high level of energy consumption and maintenance requirements became considerable weakness of the machine once the elegant and simple technology of the whirlpool was realized. The hot wort centrifuge is rarely chosen for new schemes.

Wort Cooling Stage

The objective of wort cooling is to receive the wort from the hot wort separation stage and, by the application of heat transfer, cool it to a specific temperature of 9–10°C for bottom-fermented beer and to 15–18°C for top-fermented beers. The wort must be aerated to a specific saturation to allow the cooled and aerated wort to be passed to the next stage of fermentation. The objective should be achieved with the highest efficiency to ensure that optimum wort clarity is presented to fermentation.

The wort must be cooled as quickly as possible to a level tolerated by the yeast, and air/oxygen must be injected for the yeast to achieve rapid fermentation. Aeration is most commonly performed on the cold side of the wort cooler.

For wort cooling application, plate heat exchanger has achieved total dominance and now represents the state of art equipment for highly efficient process cooling (heat transfer).

Embossing the heat exchange surfaces on both the wort and coolant side enhances the heat transfer coefficients by creating turbulence. The plate materials are usually one of the higher alloyed austenitic stainless steels and are relatively thin to maximize the heat transfer.

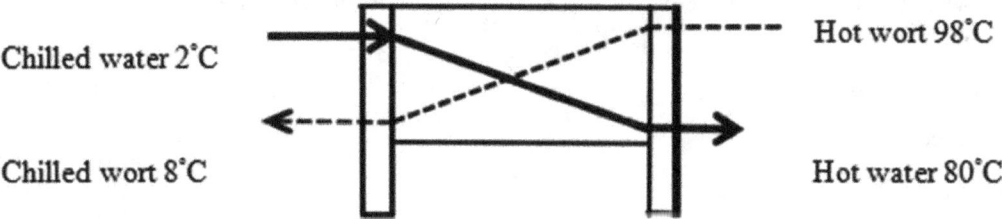

Chilled water 2°C · Hot wort 98°C · Chilled wort 8°C · Hot water 80°C

Figure 7.2: Single Stage Wort Cooler

Various solids and proteins molecules, which flocculate as cold break and are considered important to the yeast, haze and head retention factors later in the process, must not be disrupted by too much shear stress being put on the wort. The lower the value, the more frequent the CIP requirement—so effluent considerations must be considered and balanced as part of the mandatory requirement to maintain sterility and wort quality.

The ideal requirements to be achieved by the plate and frame wort cooler are:

- The heat exchange liquids should be in counter-current flow
- The velocity of the liquids over the surface should be high to promote a high Reynolds Number and a high forced convection heat transfer coefficient
- The wort constituents must not be compromised by a high degree of shear force
- The heat exchanger surfaces should be kept clean with minimal progressive fouling allowed during the process stage
- The heat exchange surface must be non-toxic or tainting and be easily cleaned and sanitized, using hot dilute acid or alkaline solutions

- The heat exchanger plates should have a high thermal conductivity coefficient and be as thin as practicable, consistent with the working pressures and the requirement to promote a high wall-conduction coefficient value

Wort Aeration

In order for the yeast to complete fermentation consistently, it needs to be first stimulated into a growth phase. The normal practice is to provide sufficient oxygen in the wort for the yeast to multiply three- to six-fold at the start of fermentation. A working rule-of-thumb is *1 mg/l* oxygen for each Plato to be attenuated.

Wort aeration can be achieved on the hot wort or cold wort side of the exchanger. Aeration on the hot side can afford a degree of sterility and additional turbulence. However, aeration of hot wort does lead to oxidation of wort constituents (polyphenols) and an increase in color and is therefore generally avoided.

The solubility of air and oxygen into the solution is significantly increased in colder wort, and the use of sterile gases and filtration cartridges favors the technique of aerating on the cold wort side.

Air injection is usually preferred as the safer, more prudent, choice over oxygen to avoid over saturating and the potential for excessive yeast growth.

In order to achieve saturation, the injected air/oxygen must be presented into the wort as very fine bubbles that must be kept in contact long enough with turbulent conditions and pressure to allow saturation to occur. In practice, the optimum is rarely achieved, and larger amounts than the theoretical injection rate are frequently found.

The air/oxygen will dissolve to varying degrees, depending particularly on the method of injection/dispersion into the wort. More finely dispersed bubbles dissolve more efficiently and available alternative techniques are:

- Sintered candle
- Wort line venturi
- In-line static mixer

Brewhouse Factors Influencing Beer Quality

Many factors influencing beer quality have their origins in the brewhouse. Good brewing practices should produce consistent wort that contains:

a. The essential yeast nutrients for vigorous fermentation
b. The fermentable extract required for the specified alcohol content
c. The correct level of hop flavor and bitterness for the desired flavor and bitterness
d. Absence of off-flavor precursors

It is important to analyze the cooled wort and monitor several critical following parameters in order to evaluate the performance of the brewhouse (these parameters represent a minimum level of testing).

- *Plato:* A measurement of dissolved solids that will determine the efficiency of malt extraction and its consistency from one to another. This measurement can also track evaporation during wort boiling which eliminates undesirable flavor-active compounds. Significant variation in Plato values can have a negative influence on fermentations.

- *pH:* An indicator of process consistency and of possible contamination with the cleaning solution, which will negatively affect quality. Wort pH values are characteristic of the recipe and will be uniform from brew to brew within a beer brand. Significant variation of pH from the norm will affect fermentation, beer flavor, and flavor stability.

- *Sodium:* Sodium has its origins in the process water and brewing ingredients. It affects flavor and is another indicator of contamination with the cleaning solution. Some brewers add sodium chloride to the wort for flavor effects. In general, the sodium content of wort should be consistent for each brand of beer.

- *Iso-alpha acids:* The concentration of iso-alpha acids in the wort parallels the quantity of hops added to the wort kettle. Loss during the fermentation renders the wort value as only a base estimate of the iso-alpha acids content of the beer. However, it will be consistent among brews among the same brand and, therefore, is a good indicator of process control and hop flavor consistency.

Other important factors:

- *Protein content* is derived from the malt-soluble protein and should be uniform when the quality of malt used and the mashing operation are consistent.

- *Free amino nitrogen (FAN)* is important for yeast nutrient. This will also be uniform when the degree of malt modification and the mash temperature schedule are uniform.

- *Zinc content* is very important for yeast nutrient and vigorous fermentation. The concentration required is 0.1–0.3 mg/l.

- *Calcium content* is also important for yeast nutrient, yeast flocculation at the end of fermentation, and precipitation of oxalate during maturation.

- *Carbohydrate profile* will provide information on the conversion of starch in the mash and on the usage of adjunct. This should also be consistent from brew to brew within the brand, showing content of fermentable sugars going into fermentation.

- *Dimethyl sulfide* measurement will tell whether wort evaporation has eliminated this off-flavor component from wort and consequently from beer. It has been found that dimethyl sulfide will form in the wort if the time in the hot wort settling tank (whirlpool) is extended.

- *Sulfate/chloride* These anions affect the beer's flavor—sulfate imparts harshness, chloride contributes fullness. Their concentrations will usually be uniform within a brand from a particular brewery. In addition to the minerals correction, corrections can come from the water, malt and adjuncts.

- *Polyphenol content* is in direct proportion to the quantity of malt, type of hop, and alkalinity of the sparge water in lautering. The concentration for a particular beer brand should be uniform from brew to brew when sparge water alkalinity is uniform.

- *β-Glucans* are derived from malt. High levels in wort increase viscosity and have a negative effect on lautering as well as on beer filtration. High concentrations are avoided by proper malt modification and mash temperature control.

Monitoring of these factors will greatly assist in identifying trends that can lead to problems in product quality. Monitoring of quality is best done through a quality system, the extent of which depends on the complexity of the operation and the available resources.

8

Fermentation

BEER FERMENTATION IS the process by which fermentable carbohydrates are converted by yeast metabolism into alcohol, carbon dioxide, and numerous other by-products. It is these by-products that have a considerable effect on the taste, aroma, and other properties that characterize the style of beer.

An understanding of the influence of yeast metabolism and growth on beer flavor will help the brewer in making decisions about process conditions and their influence on beer quality. Yeast pitching rate, temperature, pressure, agitation, dissolved oxygen, tank design, wort volume and so on, can affect fermentation both directly and interactively.

Lager Fermentation

The production of lager beers utilizes a bottom fermenting strain of the yeast *Saccharomyces uvarum (also known as Saccharomyces carlsbergensis)*. This yeast is called bottom fermenting because the yeast tends to settle to the bottom of the fermentation tank toward the end of fermentation. Brewing strains of *S. uvarum* are selected for the flavor characteristics and for other desirable fermentation characteristics, they impart to the beers. Some of these characteristics are:

- Suitable flocculation properties
- Genetic stability
- Suitable attenuation of wort carbohydrates
- Acceptable foam characteristics in the beer
- Interaction with fining materials for effective filtration

Fermentation Vessels

Modern lager fermentation usually uses vertical cylindro-conical tanks, such as uni-tank, and gradually gained acceptance because of achieving similar flavor profiles compared to traditional fermentation systems. The tanks could be used for fermentation and lagering (maturation) because the yeast can be removed from the cone, leaving the beer undisturbed in the tank. A cone angle of about 70° is best suited for yeast settling and removal. The advantages of using cylindro-conical tanks are:

- Yeast is kept in suspension better by natural agitation caused by rising CO_2 bubbles resulting in faster fermentation
- Cooling is more easily controlled
- CO_2 collection is facilitated
- Removal of yeast from the conical base is readily done
- CIP system makes tank cleaning easier
- Tanks can handle modest counter pressure, if required

Vertical cylindro-conical tanks require from fifteen to twenty-five percent of the tank volume as headspace to allow for foaming during fermentation. The amount of foam generated depends mainly on those process conditions that influence fermentation rate, and on factors that stabilize foam. The concentration of hop constituent in wort is a major factor in the stability of foam during fermentation.

Characteristics of Fermentation

Traditional lager fermentations are conducted at temperatures ranging from about 9–14°C. The fermentation temperature in a fermenter is usually allowed to rise spontaneously to a few degrees above the pitching temperature. Because of the heat generated during fermentation (about 140 kcal/kg extract fermented), cooling is required to limit the temperature rise. The duration of fermentation traditionally ranged from eight to twenty days, but with the higher temperatures often used in modern practice, fermentation time may be reduced to about seven days.

During the first six to ten hours of fermentation, the yeast consumes all of the dissolved oxygen. There is no detectable uptake of glucose during this time. At about eight to sixteen hours, the first signs of active fermentation appear when CO_2 bubbles are formed and thin foam or head is apparent.

The budding of cells can be observed within twenty-four hours. The temperature, if uncontrolled, begins to rise due to heat generated by fermentation. Within twenty-four to forty-eight hours, the rates of yeast growth and carbohydrate assimilation reach their maxima. The rate of carbohydrate assimilation and the progress of fermentation can be measured by the CO_2 evolution rate.

The pH falls as organic acids are produced and buffering compounds (basic amino acids and primary phosphates) are consumed. The minimum pH attained during fermentation is a function of three factors:

- Wort pH
- Wort buffering capacity
- Amount of yeast growth during fermentation

Lower beer pH is associated with a lower wort pH, lower wort buffering capacity, and increased yeast growth. The pH reaches a minimum of 3.8–4.4 before rising slightly toward the end of fermentation. The lowered pH inhibits bacterial spoilage during fermentation.

High krausen, (i.e.) maximum foam head, occurs at the time of the maximum rate of CO_2 and heat generation. The maximum fermentation rate (activity) also corresponds to the maximum decline in specific gravity. Because foam is stabilized by proteinaceous compounds and isohumulones from hops, a 'krausen ring' of precipitated foam components adheres to the fermenter walls above the liquid as the foam head subsides toward the end of fermentation.

Glucose and fructose are consumed first and, as the glucose concentration diminishes, the enzyme systems required for assimilating maltose are synthesized and the yeast begins to utilize maltose and maltotriose. The production of ethanol and other fusel alcohols generally follows the consumption of carbohydrates. There is usually a delay of about one day after the beginning of fusel alcohol production before the production of esters is observed. The fermentation rate begins to slow, the krausen begins to fall, and heat generation diminishes.

During growth, yeast produces two a-acetohydroxy acids, which are excreted into the wort and converted into diacetyl and 2,3-pentanedione. Diacetyl and 2,3-pentanedione are collectively referred to as vicinal diketones or VDKs. Yeast assimilates VDKs toward the end of fermentation. When fermentation ceases due to lack of fermentable carbohydrates, the amount of VDK precursors may still present a potential flavor defect. It is, therefore, important to allow enough time for the total amount of the VDKs and their precursors to be reduced below their flavor threshold or to an acceptable concentration before complete removal of yeast. The reduction of VDKs has traditionally been treated as a maturation problem.

When fermentation reaches the attenuation limit and the level of VDKs and their precursors are low enough, the yeast is separated from the beer that is then transferred to storage tank for maturation (lagering). Yeast can be separated by decanting the beer or, more effectively, by centrifuging or filtering it. Cooling the fermenter before transfer will help the yeast to settle and aid the transfer process by reducing yeast carryover. If ruh storage is desired, some yeast is intentionally carried into aging. In a uni-tank operation, the yeast is removed from the bottom of the tank, while the beer remains behind for the maturation (lagering).

If a secondary fermentation is desired, the transfer from the fermenter will be done with some fermentable extract still remaining (1–4 Plato). More yeast carryover is permitted so that the cell concentration is $1–4 \times 10^6$ cells/ml.

Ale Fermentation

In terms of yeast biochemistry during fermentation, there are few differences between ales and lagers. Major differences occur in the processing conditions, traditional fermentation vessels, and methods of yeast recovery. Modern practices are lessening the differences in vessel use and yeast recovery.

Traditionally, ale fermentations use the top-cropping yeast strain *Saccharomyces cerevisiae*. This strain of yeast rises to the top of the beer toward the end of fermentation because the yeast flocs entrap CO_2, making them buoyant. However, with deep cylindro-conical vessels, brewers may use a bottom-cropping *S. cerevisiae* strain to make ale.

The fermenter may be of the open type, which allows easy removal of the yeast crop by skimming or suction. The many fermenters developed in the UK were designed with yeast skimming as an important consideration. Closed vessels have become more common, and suction devices are designed to be used in closed tanks. An alternative is to drop the ale out of the bottom of the tank, leaving the yeast crop behind.

Ale wort is pitched at a higher temperature (about 15°C) than in lager fermentations and the temperature is allowed to rise to about 20°C or higher. Early in fermentation while the fermentation is active (twenty-four to thirty-six hours), the fermenting beer may be transferred to another vessel. This helps mix and aerate the wort. By pump circulation, the mixture may be 'roused' instead. The higher temperature and extra aeration causes ale fermentations to be completed in much less time than lagers (≤3 days). Secondary fermentation follows a course similar to that used with lagers.

As additional foam head is formed during fermentation, and yeast crop removal may be carried out in steps. The yeast retained for re-pitching must be carefully selected from these separate skimming operations because the yeast collected at different times may have different attenuation characteristics. For example, the yeast crop formed initially may have less ability to fully attenuate the wort.

Short fermentation time, high temperature, and different yeast strains all contribute to a product that has a different balance of various flavor compounds such as fusel alcohols and esters. These give ales a distinctly different flavor from lagers.

The main fermentation can be divided into 4 phases.

1. Lag Phase

This stage takes from half-a-day to two days depending on yeast pitching rate, yeast viability, yeast vitality, wort temperature and wort aeration rate before the fermentation start. The yeast cells need time to adapt to new surroundings and produce some enzymes that are necessary for transporting substances into the cell membrane.

2. Log (accelerating) Phase

In this stage, the yeast cell is increasing up to the maximum (normally three to five times the pitching rate). The beer gravity will be dropped about two to three percent Plato per day. Because of yeast metabolism, when the beer temperature reaches the maximum, it is time to implement the beer cooling by maintaining consistent temperature. One will see the foam on surface of the beer—this form consists of hop resins, color, tannin and cold break.

3. Stationary Phase

The beer gravity will be dropped by about three to five percent Plato per day depending on the beer's temperature, top pressure and yeast vitality and viability. The beer foam becomes dark because of the settling of the hop resin, color substances and tannin.

4. Decline Phase

In this stage, the beer fermentation rate starts to decrease and becomes constant (end fermentation) because of the reduced content of fermentable sugar in the beer. The beer temperature will cool down to maturation temperature set point—for example, 7°C until the VDK is less than 0.15 ppm.

Laboratory Analyses during Fermentation

Several laboratory analyses are necessary to follow the progress of fermentation, determine its completion, and monitor it for abnormalities. The simplest measurement, and also one of the most useful, is the specific gravity. After removal of the yeast, specific gravity measurement can be estimated with a hydrometer or, more accurately, by measurement of mass and volume. As sugars are consumed and alcohol is produced, the specific gravity falls. When the specific gravity stops declining, it indicates the end of carbohydrate fermentation. In contrast to wort, in which the specific gravity corresponds directly to the solids concentration in wort, the specific gravity does not quantify the true solids (*real extract*) remaining in beer. Alcohol is less dense than water so the specific gravity is only an indication of the *apparent extract* left in the wort.

Apparent extract indicates the actual reading of the specific gravity of the beer, which is a combination of the extract content and the produced alcohol. The alcohol causes a thinning effect, giving a lower reading of the actual extract content—hence the terminology 'apparent'.

The apparent extract is always lower than the real extract but is still a useful indicator of fermentation progress and can be used as a beer specification. *Present gravity* is sometimes used to describe the gravity during fermentation. *Real extract*, which expresses true solids concentration in beer, can be determined from dealcoholized beer, or from specific gravity and refractive index. Both real and apparent extracts are expressed as Plato. Modern instruments can be used to easily estimate the various parameters described here. A measure of how much carbohydrate was fermented is the real degree of fermentation (RDF), which measures the percent of extract that was fermented.

Brewers measure the real degree to which sugar in cold wort has been fermented into alcohol in beer with the term degree of fermentation (RDF). A sweet beer has more residual sugar (unfermentable sugar) and lower degree of fermentation. Like the calculation of extract there are several methods of expressing degree of fermentation. Real degree of fermentation is an expression for the fraction of extract originally present in cold wort, which has been transformed into alcohol and CO_2. It is calculated from the variables Alcohol (A) and Real Extract (RE) as determined by the beer analysis in the following equation:

$$RDF\ (\%) = \frac{(O - E)}{O} \times \frac{1}{1 - (0.005161 \times E)} \times 100$$

Where,

 O = *Wort original gravity (Plato)*

 E = *Real extract (Plato).*

The second term accounts for weight loss due to the increase in yeast mass and loss of CO_2. The second term permits the RDF to be used with high-gravity brewing because it is not affected by dilution. An RDF of sixty to seventy percent is normal for lager beers.

Cell counts will indicate abnormal growth patterns during fermentation, which may result from improper process conditions, wort composition, or pitching yeast. Dead cells can be determined by staining them with methylene blue and counting them with a hemacytometer. The method suffers from subjectivity due to variation in the amount of methylene blue that the dead cells absorb. Analysts should standardize their techniques to reduce between-analyst variability. Checks for dead cells can be made during fermentation, at the end of fermentation, or in the yeast brinks after collection. Dead cells tend to remain in suspension and, on a percentage basis, may appear to be increasing toward the end of fermentation when, in fact, they are not.

More extensive laboratory analyses are possible, which help the brewer diagnose aberrant conditions and maintain quality products. For example, SO_2, alcohol, and VDKs can be monitored. Estimations of diacetyl and 2,3- pentanedione (VDKs) are commonly used by brewers as a specification before transferring beer or releasing beer from maturation.

Factors Affecting Fermentation

Part of the flavor profile of beer is based on yeast metabolism during fermentation. Consequently, wort composition and process conditions affecting the fermentation performance of yeast will also affect beer quality. These factors are so closely interrelated, it is often difficult to clearly single out the influence that any one factor exerts on fermentation or product quality. Attempting to alter one process parameter to influence an outcome almost always causes other, perhaps unwanted, outcomes.

• Yeast Strain and Condition

The yeast strain itself is a major contributor to the flavor character of beer. Each yeast strain may perform differently under a given set of fermentation conditions and the brewer must consider the flavor profile produced by a strain (sulfur compounds, esters, fusel alcohols, and so on). The choice of yeast strain depends on characteristics considered important—the relative importance of oxygen requirements, cropping methods, attenuation limits, fermentation rate, and so on.

The flocculation characteristics of a strain are important. The ideal yeast would settle rapidly as the wort reaches limit attenuation. A slowly flocculating strain would have too much yeast in suspension at the end of fermentation and make beer separation more difficult. A rapidly flocculating strain may not ferment fully. On the other hand, if secondary fermentation were intended, a more slowly flocculating strain, leaving a greater number of cells in suspension at transfer would be desirable.

- Pitching Rate and Yeast Growth

The pitching rate has an influence on the fermentation rate. The fermentation rate depends on the temperature and cell population. Higher pitching rates give shorter lag phases, higher maximum fermentation rates and shorter times to complete carbohydrate fermentation. A typical pitching rate of about 10-15 × 10⁶ cells/ml is seen for normal gravity lager wort (about 12°Plato). The precise concentration chosen depends on the yeast strain, fermentation characteristics desired, and so on.

Because the growth of new cells is limited by the oxygen supply in the wort, the increase in cell growth is nearly independent of initial yeast concentration under fixed, dissolved oxygen conditions. Compared to low pitching rates, higher pitching rates will lead to a lower number of cell doublings. At some point, yeast growth does depend on pitching rate. Insufficient growth will lead to slower and, possibly, stuck fermentations. Poor yeast growth can also lead to high SO_2 concentrations.

Variations in cell growth and therefore variations in fermentation rate at fixed pitching rates and dissolved oxygen are possible if other yeast growth factors in wort vary (such as sterols or zinc). Different yeast growth patterns will influence the flavor of the beer by changing the proportions of volatile flavor compounds. For this reason, consistent yeast growth in all fermentations should be a primary objective.

- Temperature

To minimize the consequences early in fermentation from process variations, fermentation temperature control is necessary for consistent yeast growth. Fermentation rates will increase with temperature by increasing the rate of yeast metabolism giving higher specific fermentation rates. Higher temperatures will give a faster conversion of VDK precursors and shorten the time required to reduce potential VDK off-flavors. Faster fermentation rates raise the peak demand for cooling. The quantitative influence of a temperature change will be different for each biochemical reaction, changing the balance of flavor compounds.

- Oxygen

The dissolved oxygen concentration in the wort will influence yeast activity in fermentation because yeast requires oxygen to produce essential compounds for new yeast cells. In normal wort, oxygen is the limiting nutrient of yeast growth. Wort saturated with air provides enough oxygen for proper yeast growth. However, consistent control over the dissolved oxygen in wort is essential for uniform growth in each fermentation and consistent production of flavor compounds.

- Zinc

Zinc, as Zn^+ ion is required as a cofactor in enzymatic reactions within the cell and therefore is required for growth. The addition of zinc has been clearly linked to increased fermentation rates. A zinc concentration of 0.3 ppm (mg/l) should be adequate in most cases. Yeast will utilize nearly all zinc present in wort. It has been found that concentrations of zinc above 1 ppm (mg/l) can inhibit yeast activity.

- Trub Carryover

Trub in the fermenter can influence the fermentation rate and filtration performance. Because of the amorphous nature of trub and settling characteristics in the hot wort tank, however, it may be difficult to accurately control the quality and quantity of trub carried into the fermenter. Trub contains a mixture of substances that can be beneficial to yeast performance—zinc, lipids and sterols, and other minerals and nutrients essential for yeast growth. It also provides nucleation sites for bubble formation, increasing the fermentation rate by reducing dissolved CO_2 and encouraging yeast growth, and increasing turbulence from rising bubbles.

- Fermenter Geometry

Vessel geometry plays an important role in fermentation. The depth of the tank is important because the hydrostatic head affects CO_2 bubble formation hence, dissolved CO_2 concentration and yeast growth and mixing. It has been shown that yeast growth and flavor production were reduced under higher dissolved CO_2 concentrations, which would result from greater hydrostatic heads. The differences in fermentation characteristics between horizontal fermentation and vertical tanks are the result of natural agitation from convection currents and rising CO_2 bubbles, and a dissolved CO_2 difference. As the height to diameter ratio increases, so do natural agitation and fermentation rate. Generally, horizontal tanks give slower fermentations and produce more estery beers than vertical tanks.

Yeast collection procedures for horizontal and vertical tanks differ significantly. Compared to vertical tanks, horizontal tanks generally require manual removal of yeast.

- Interrelationships

A component of the art of brewing is to find the proper combination of process conditions that produce the optimum product. In most cases, the beer flavor profile will be affected, as the main impact of a change is on yeast growth. For example, additional unsaturated fatty acids and sterols in trub may lessen the requirement for oxygen. A lowered pitch rate may be compensated by higher temperature in terms of time to limit attenuation.

The requirements of a particular yeast strain for specific nutrients (wort composition) and sensitivity to different environments during processing will dictate process conditions so that the strain performs well.

Yeast handling must be carefully monitored because the condition of the yeast at the time of pitching has been shown to influence fermentation. Yeast stored under extreme conditions may not respond normally, having reduced its glycogen in response to stress. Poor aeration, zinc deficiencies, or residuals from previous fermentations may cause an excessive lag phase or incomplete attenuation.

Abnormal Fermentations

A. Symptoms

Fermentation is a natural process that occurs freely in nature. As a result, industrial beer fermentations tolerate wide variations in process conditions. Although abnormal fermentations are rare, 'stuck' or 'hung' fermentations will occur even in the most modern breweries utilizing elaborate process control and monitoring equipment. Symptoms of a stuck fermentation may be a long lag phase accompanied by a very slow fermentation rate, followed by no fermentation activity at all. In other cases, after a normal lag phase active fermentation may simply stop before all fermentable carbohydrates are consumed.

B. Causes

1. Process Variations

The pitching rate, yeast viability, level of aeration, and wort fermentability should be the first areas to be investigated—these process conditions are among the easiest to check. Usually, a check of meter calibration and settings, time-temperature records, yeast cell counts (including dead cells), and a microbiological examination will point to the problem. Insufficient wort aeration and low pitching rates are common causes of abnormal fermentations. Successive fermentations in nutrient-deficient worts can gradually lead to deterioration of the yeast and slow fermentations. Checks of yeast cell counts in the fermenter may show early flocculation of yeast or its inability to completely assimilate maltose or maltotriose. A positive starch test will indicate a brewhouse conversion problem.

2. Wort Nutrient Deficiencies

A more difficult situation occurs if there is a deficiency in wort nutrient. If slow or stuck fermentations are a continuing problem, wort deficiencies may be the cause. Under normal conditions, nutrients for yeast growth are in excess. However, if the yeast has been handled poorly, thereby increasing its requirement for a particular nutrient, the wort may not be able to supply it. Also, changes in malt or brewhouse processes or raw materials may change wort composition and produce a nutrient deficiency.

The most common deficiencies are oxygen, zinc, biotin, unsaturated fatty acids and sterols. As discussed earlier, yeast in anaerobic fermentation has a requirement for oxygen, unsaturated fatty acids, and sterols. These requirements are interrelated. Usually, sufficient oxygen will reduce the need for unsaturated fatty acids and sterols in wort and vice versa. Biotin is obtained from malt during mashing and is a growth factor required by most brewing yeasts.

When using a high percentage of adjuncts, another possible nutrient deficiency is reduced nitrogen content in the wort. Insufficient amino acids will hinder proper yeast growth. In worts with high glucose concentrations relative to the other sugars, 'stuck' fermentations may occur from a condition sometimes referred to as glucose repression. In this case, the presence of high concentrations of glucose prevents the yeast from synthesizing the enzymes needed to assimilate the other sugars. A similar condition may arise if a high concentration of fructose is present causing 'fructose block'. These situations are more likely in worts for low-calorie (low-carbohydrate) beers because the adjunct may be solely glucose.

3. Yeast Changes

Another possibility is that yeast performance has changed during successive re-pitchings. Such changes may be caused by a build-up in yeast cells of toxic amounts of normal nutrients; adsorption of hop and trub compounds on the surface of the yeast; a mutation that affects sugar utilization; or a contaminating killer yeast strain.

C. Treatments

If a process variation was discovered to cause a stuck fermentation, appropriate action is obvious. Contamination of the yeast requires careful investigation of possible sources and corrective action. *Wort nutrients can be increased by adding yeast extracts or yeast food.* These preparations have a variety of nutrients beneficial to yeast growth. In the case of persistent fermentation problems because of wort nutrient deficiencies, more laboratory analyses and study will be required to pinpoint which compound is needed. Yeast deterioration can be eliminated by using a freshly propagated yeast batch. Increasing trub carryover may have a beneficial effect on fermentation.

If fermentation is 'stuck', one may want to recover the product. The addition of ten to twenty percent volume of actively fermenting beer at high krausen may help; in severe cases, the blend may need to go up to fifty percent. Another, although more risky, the remedy is to aerate the fermentation. This may activate the yeast, but it introduces the possibility of contamination, or may cause oxidation reactions and off-flavors. If fermentable sugars in the wort are too low, the addition of an amylolytic enzyme to the fermenter may substitute for a poor mashing process.

Beer Transfer and Yeast Separation

Most brewers periodically propagate their yeast from pure cultures. However, generally, there is no need to propagate yeast for batch-wise fermentation, as there is sufficient yeast available from production fermentations to supply nearly all of the needs of a brewery. There is a four- to six-fold increase in yeast concentration during fermentation. Aside from the need to remove most of this yeast from beer prior to aging or secondary fermentation, yeast recovery for reuse in subsequent fermentations is an important process in any brewery.

A. Yeast Cropping Considerations

Yeast strains may have an important influence on the method chosen for separating yeast from beer. More powdery strains are best removed by centrifugation or filtration, while highly flocculent strains are more efficiently separated by sedimentation. Moderately flocculent strains, which allow more yeast carryover during transfer, are used if a secondary fermentation or ruh storage is desired. The formation of yeast flocs or clumps is usually aided by the presence of calcium ions, low concentrations of fermentable sugars, low pH, and low temperature. These are the conditions at the end of fermentation. Higher amounts of zinc assimilated during fermentation also aid flocculation. Ale strains have been classified into four groups based on the requirement for an inducer for flocculation. It is believed that certain amino acid residues in peptides are the inducers.

The geometry of fermenting vessels also plays a role and, in fact, usually determines the separation method. Shallow vessels of large surface area are conducive to relatively complete sedimentation of highly flocculent yeast. Cylindro-conical vessels are better for more efficient yeast separation of less flocculent strains, as the bottom fermenting yeast collects in the conical bottom. The larger volume cylindrical tanks are even useful for top fermentations.

B. Methods of Cropping

In traditional top-cropping ale fermentations in open vessels, the yeast crop is skimmed by various devices. A parachute device (inverted funnel) and suction or similar suitable system is common. This technique has also been used successfully with closed fermenters. Other techniques, used principally in the UK, rely on specially designed fermenters that aid in collection of the yeast.

In lager fermentations using horizontal tanks, the beer is simply drawn off for lagering, leaving the yeast behind. The yeast is then collected manually from the vessel floor. This yeast contains other, mostly proteinaceous sediment from the fermenting beer, and separation of the yeast from the debris is difficult. Modern cylindro-conical tanks allow improved separation and collection strategies. Temperature reduction at the end of primary fermentation aids sedimentation of yeast into the cone for easy beer transfer and yeast collection. Most of the yeast can be collected from the conical bottom before the beer, which contains some yeast in suspension, is removed from the fermenter. The beer can then be transferred to ruh storage with the residual yeast present, or the yeast can be removed by passing the beer through a centrifuge. Alternatively, the beer can be transferred first, leaving the yeast layer undisturbed, for more efficient removal. In large vertical fermenters in which bottom fermenting strains are used, the trub will settle with the early flocculating layer. Removal of this layer to waste will eliminate most trub particles in the remaining yeast crop. The ease of collection in cylindro-conical tanks makes it advantageous to use bottom fermenting strains of S. cerevisiae for ale production.

Because yeast will flocculate continually toward the end of fermentation, there will be a somewhat stratified yeast crop with varying flocculation characteristics. Collection of the bottommost layer in bottom fermenting strains or the topmost layer in top-fermenting strains will harvest the most flocculent yeast, while other layers will contain the least flocculent yeast. Re-pitching yeast with undesirable flocculation characteristics may lead to fermentation problems. Separating yeast from green beer by skimming the yeast or decanting the beer will not leave the beer totally free of yeast. If a secondary fermentation or ruh storage is used, some yeast is needed in aging, and yeast carryover is not a problem. If no secondary fermentation is used, then more efficient transfer is advantageous. A better separation of yeast from beer can be accomplished by using centrifuges.

C. Centrifugation

With the introduction of centrifuges in the brewery for yeast separation, the need for shallow tanks was eliminated, and the use of tall, space saving tanks was promoted. The use of centrifuges for clarification has become more common with the development of equipment that can be used successfully with beer. For example, modern centrifuges have hermetic seals that:

- Exclude air during their operation
- Minimize temperature increases
- Reduce turbulent flow

There are two principal ways to use centrifuges for beer separation after fermentation:

1. To use the centrifuge to separate the yeast crop from the entire fermenter—this accomplishes beer transfer and yeast collection in one step
2. To use a centrifuge to clarify the beer after the yeast has been separated by decanting or skimming for re-pitching

This two-step procedure is generally used when the collected yeast will be used for re-pitching, because centrifugation can be stressful to the yeast cell.

Centrifugation can heat the beer and yeast streams. It is necessary to have a cooler for the beer to bring it to maturation/lagering temperature. If the yeast collected by centrifuge is to be used for re-pitching, a cooler should be used to reduce the temperature of the yeast to brink temperature as quickly as possible.

The principle of centrifugation is based on Stokes' law governing the rate of settling of particles. A greater settling rate occurs with:

- A lower viscosity liquid
- Larger diameter particles
- Greater difference in density between the particle and the liquid.

None of these factors is usually controllable. However, the settling rate is increased by the considerable force created by the centrifuge (up to 8,000 times the force of gravity). A shorter settling distance, which is a common feature of modern centrifuges, also speeds sedimentation.

There are two basic types of centrifuges used in brewery applications. They may be classified according to the load of solids that they handle.

- A dewatering or decanting centrifuge is a screw-type conveyor, generally horizontal, that handles larger and more fibrous particles in liquids of up to about sixty percent solids. The centripetal force moves the particles to the outer surface of the cylindro-conical shell, where they are conveyed by the screw to the discharge. It is used mainly for brewery effluents, to recover liquor containing extract from brewer's grains after lautering, and to recover beer from tank bottoms containing yeast.
- The second and most common type of centrifuge, a clarifying centrifuge, is a disk-stack bowl type in a vertical configuration. Clarifying centrifuges use either intermittent or continuous ejection of solids. Continuous ejection models are referred to as nozzle centrifuges. The disk-stack centrifuges can handle liquids with up to about thirty percent solids and are ideal for brewery applications, as they are self-cleaning, air tight, and have CIP systems. In a disk-stack bowl centrifuge, the centripetal force moves the particles to the outside of the bowl, where solids are removed intermittently by rapid openings of the bowl, or continuously through a nozzle. Manufacturers produce a large variety of models suitable for a broad range of brewery applications.

Clarifying centrifuges perform best when they receive a feed stream with a uniform concentration of solids. Therefore, they are most efficient for use with:

a. Beer that has already been cropped—for example, ruh beer that contains only a low concentration of yeast
b. Fermentations using powdery yeast strains

The use of centrifuges in the brewery was initially viewed with misgivings, but the advantages and disadvantages are now reasonably clear.

Advantages of centrifuges are:

- Rapid and efficient clarification before further filtration steps
- Consistent clarity
- Equipment can be sterilized
- Filter aids are not required
- Space requirements are small
- Most are self-cleaning
- Operate continuously
- Lower beer losses compared to sedimentation
- Minimal oxygen pick-up

Disadvantages of centrifuges are:

- High maintenance costs
- Beer temperature may increase
- Yeast needed for flavor maturation (VDK reduction) may be removed completely
- Mechanical break-up of large particles due to shear may increase the concentration of finer haze particles
- Increased mechanical stress or increased yeast temperature may adversely affect yeast being used for re-pitching
- Improper operation possibly leading to oxygen pick-up, and high noise level

When using centrifuges, the three most important factors for maintaining beer quality are oxygen exclusion, a microbiologically clean operation, and a minimal beer temperature rise as a result of processing. If proper cleanliness and strict operating standards are employed, beer separated from yeast by centrifuge will be of excellent quality.

Recovery of Carbon Dioxide Gas

Less than ten percent of fermentable sugars are converted into new yeast mass during fermentation. The remaining fermentable sugars are converted into approximately equal quantities of ethanol and CO_2. Depending on wort gravity, about 3.6–4.5 kg of CO_2 are produced/hl and theoretically collectible; but in practice, about only eighty percent or less is recoverable depending on fermenter geometry (headspace), fermentation rate, and efficiency of the recovery system. The uses for CO_2 are:

- Sparging beer in aging
- Counter pressure in beer storage tanks
- Purging filters and transfer lines
- Packing operations

These can usually be met with the CO_2 gas recovered from fermentation, assuming there are no losses. Major uses are carbonation (about 0.5 kg/hl for each volume CO_2 used), tank counter pressure and transfer (1 kg/hl), and packaging, especially in can operations (up to 1.5 kg/hl). The collection, purification, and liquefaction of CO_2 for reuse thus become economically attractive. With careful attention, brewers can meet their CO_2 needs with the gas recovered from fermentation.

Purity and Collection Strategies

Just as the introduction of oxygen into beer is detrimental to its flavor, the CO_2 used in processing must contain as little oxygen as possible. The oxygen in CO_2 is influenced by collection strategies from the fermenter as well as proper design and operation of a purification system.

To minimize oxygen impurities in CO_2, the collection timing must be coordinated with the fermentation cycle. *In particular, the CO_2 evolved early in fermentation must be vented until its oxygen concentration is below some specified limit, usually 0.01 percent or less.* In-line sensors are available to assist in timing of collection. The duration of venting will depend on the fermentation rate, fermenter geometry, headspace volume, and product type.

Fermenters are disconnected from the collection system toward the end of carbohydrate fermentation when CO_2 production is low. Because fermenters are added to the production stream according to a known brewing schedule, the peak CO_2 load on the collection system can be calculated. From this calculation, the collection system can be properly designed to handle the CO_2 expected from normal production.

9

Lagering & Finishing

(MATURATION AND FILTRATION, Standardization, Carbonation)

Lagering/Maturation

The maturation of the green (immature) beer to produce a stable, quality product suitable for filtration and packaging is called lagering or, alternatively, cold conditioning or cold storage.

A. Objectives of Lagering (Aging) and Finishing

Lagering refers to flavor maturation. At the end of fermentation, many undesirable flavors and aromas of a 'green' or immature beer are present. The lagering process reduces the levels of these undesirable compounds to produce a mature product.

'Finishing' refers to the production of a brilliantly clear beer after lagering that remains that way until it is consumed.

B. Component Processes

The component processes of aging and finishing are:

1. Lagering or aging
2. Clarification
3. Stabilization
4. Carbonation
5. Blending or standardization

Each process can be accomplished in a variety of ways, but each is independent and can be treated as a unit operation.

In modern practice, cold aging or lagering is the storage of beer for the purpose of flavor maturation.

Table 9.1: Unit Operations for Lagering & Finishing

Unit Operation	Purpose	Equipment & Method
Transfer	Yeast Separation	Decant beer/Centrifuge/Filter
Lagering	Flavor maturation	Some yeast present for VDK reduction/CO_2 purging
Stabilization	Protect beer from:	
	1. Oxidized flavor	Keep yeast present/Use CO_2 during transfer of beer/Add antioxidant
		Pasteurization/Sterile filtration
	2. Biological haze and off-flavors	Use proteolytic enzymes/Isinglass fining
	3. Physical haze (chill proofing)	
		Filters
Clarification (Filtration)	Removal of all suspended particles	
Carbonation	Attain proper CO_2 concentration	Traditional lagering/Pressurized fermentation/CO_2 injection
Standardization (Blending)	Uniformity of packaged beer	Blenders

After lagering, clarification is required to remove any remaining yeast and suspended particles formed during cold storage. At least one filtration step is needed before beer is suitable for packaging if a clear, brilliant beer is desired.

Stabilization refers to protecting the finished product from changes that may occur after packaging. These changes are:

- Flavor changes primarily due to oxidation.
- Non-microbiological haze caused by the formation of molecular complexes.
- Haze produced by the growth of bacteria or yeast.

Blending or standardization is the process of mixing batches of beers to achieve uniformity of flavor or analytical characteristics.

Carbonation is the process of adjusting the CO_2 concentration to a specified concentration. Carbonation by injection of CO_2 into beer is done as a replacement for the traditional raising of the CO_2 level by a cold secondary fermentation.

Brewers may combine some of these operations, or change the order in which they are carried out. The possible variations are too numerous to detail here, but most brewers, for reasons of economy and product uniformity, attempt to combine some of the unit operations.

Flavor Maturation

A. Introduction

Flavor maturation is generally considered the most significant effect of lagering. Successful flavor maturation has become more important as beers have become 'lighter' in flavor. Taste thresholds of objectionable flavors are lower in lighter beers. In heavier beers, the higher concentrations of flavorful compounds will mask some objectionable flavors and aromas.

Flavor maturation can be described in terms of individual compounds that can be detected in the wort and beer—the brewer can rely on laboratory tests, in addition to taste tests, to determine the success of maturation. Taste tests can be unreliable and should be used with the knowledge that tasters vary in their sensitivity to different flavors. Therefore, most brewers supplement tasting with chemical tests and set specification limits on objectionable flavor compounds. In process beer must meet such specifications and satisfy taste requirements before release to downstream processing.

Because most of the important compounds responsible for flavor maturation are a result of yeast metabolism, the central role of consistent yeast growth during fermentation is once again stressed.

B. Important Flavor Compounds

1. Diacetyl and 2,3-Pentanedione

Diacetyl and the homologous compound 2,3-pentanedione (collectively are called vicinal diketones or VDKs) have flavor properties that are important to the brewer. Both compounds have a buttery flavor generally considered objectionable in lighter-bodied lagers but sometimes desirable in ales and more full-bodied beers. Diacetyl has a higher flavor impact than 2,3-pentanedione. The flavor threshold of diacetyl and other flavor compounds as well, depends on the background flavor intensity of the beer but it is usually detectable at about 0.1 mg/l. It is thought that at sub-threshold levels, diacetyl contributes positively to palate fullness. Brewers may speak of diacetyl only, but both VDKs are important to maturation.

The precursor to diacetyl, α-acetolactate is produced by the yeast as it synthesizes the amino acids valine and leucine needed for protein synthesis. The α-acetolactate is transported out of the cell, where it is converted nonenzymatically to diacetyl. This step is the slowest or rate-limiting step and is accelerated by a higher temperature and lower pH. The diacetyl is subsequently re-assimilated by the yeast and reduced enzymatically to butanediol by way of acetoin. The importance of this step is that butanediol has virtually no impact on flavor. A similar series of reactions occurs for 2,3-pentanedione, the precursor of which is α-acetohydroxybutyrate.

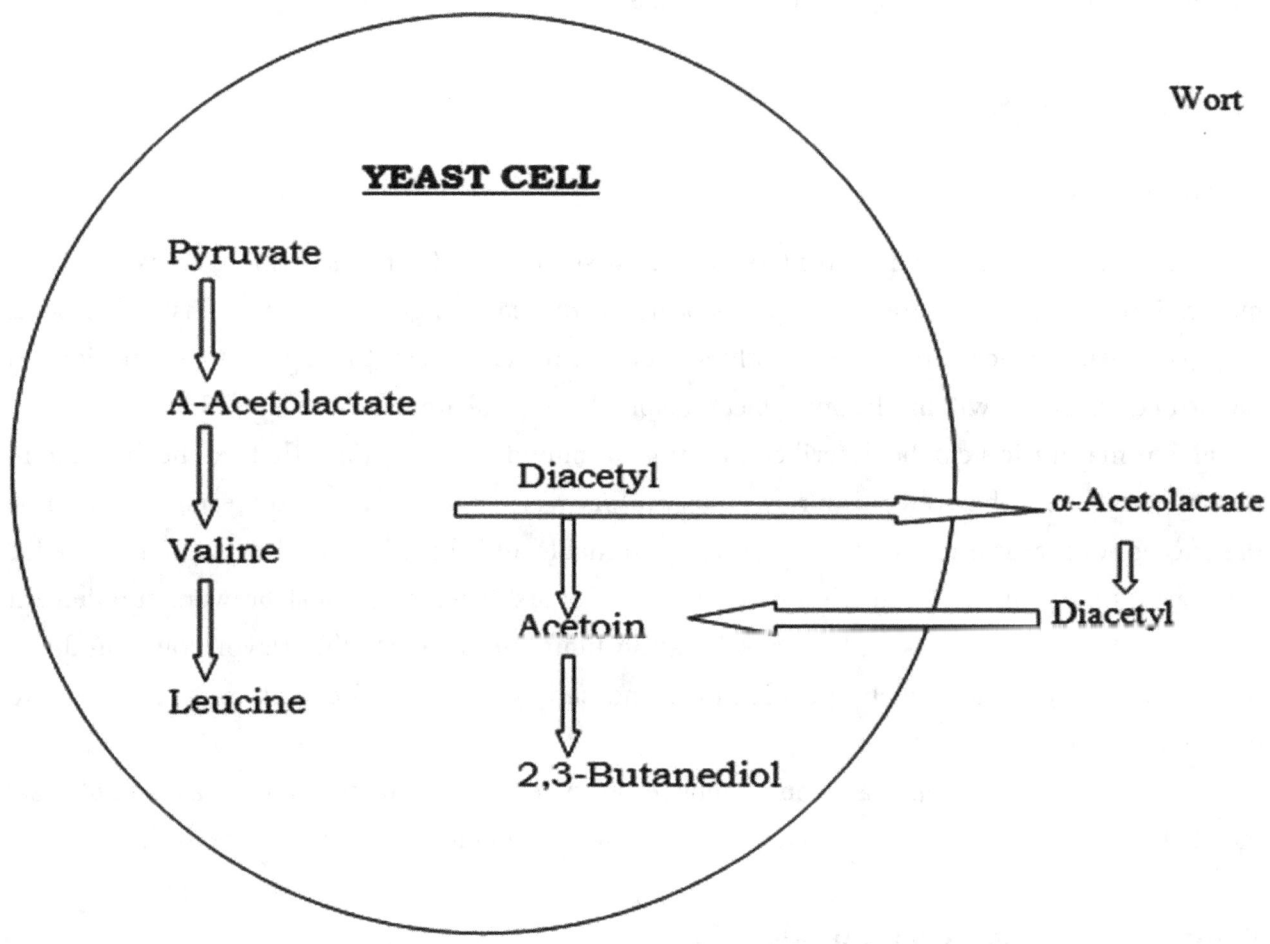

Figure 9.1: Diacetyl and the Yeast Cell

The brewer should be concerned about the concentration of precursors and whether or not yeast is present to remove VDKs when formed. The important concepts are:

- The precursors are produced as a result of yeast growth relative to the wort valine and other amino acid concentrations
- The precursors are potential flavor-active VDKs
- Conversion of precursors to VDKs is an extracellular chemical reaction that varies with temperature, pH, and so on

114

- These extracellular reactions are rate-limiting in the conversion of precursors and removal of VDKs from beer
- Yeast assimilates VDKs, and therefore needs to be present to reduce the VDKs as they are formed

The production of precursors continues throughout carbohydrate fermentation (Figure-2). In reality, the profile marked diacetyl in Figure-2 includes both diacetyl and its precursor, because the chemical test generally converts the precursor; thus, the graph shows the total potential diacetyl in the final product. Because precursor conversion to diacetyl is the rate-limiting step, with yeast present in fermenting wort, the concentration of diacetyl is small compared to the precursor. Considerable potential for VDK formation remains after active fermentation because of the high concentration of precursors.

The maturation process has two objectives: the spontaneous conversion of precursors to VDKs and their removal by yeast. Thus, to hasten the conversion of precursors to VDKs, the temperature in the fermenter can be held higher (for example, about 15°C) for a period of time after the completion of carbohydrate fermentation. Once the total precursors (potential VDKs) and VDKs fall below a specified level, the temperature can be lowered to help in yeast sedimentation. However, the higher temperature may lead to other off-flavors from non-volatile yeast products or yeast autolysis.

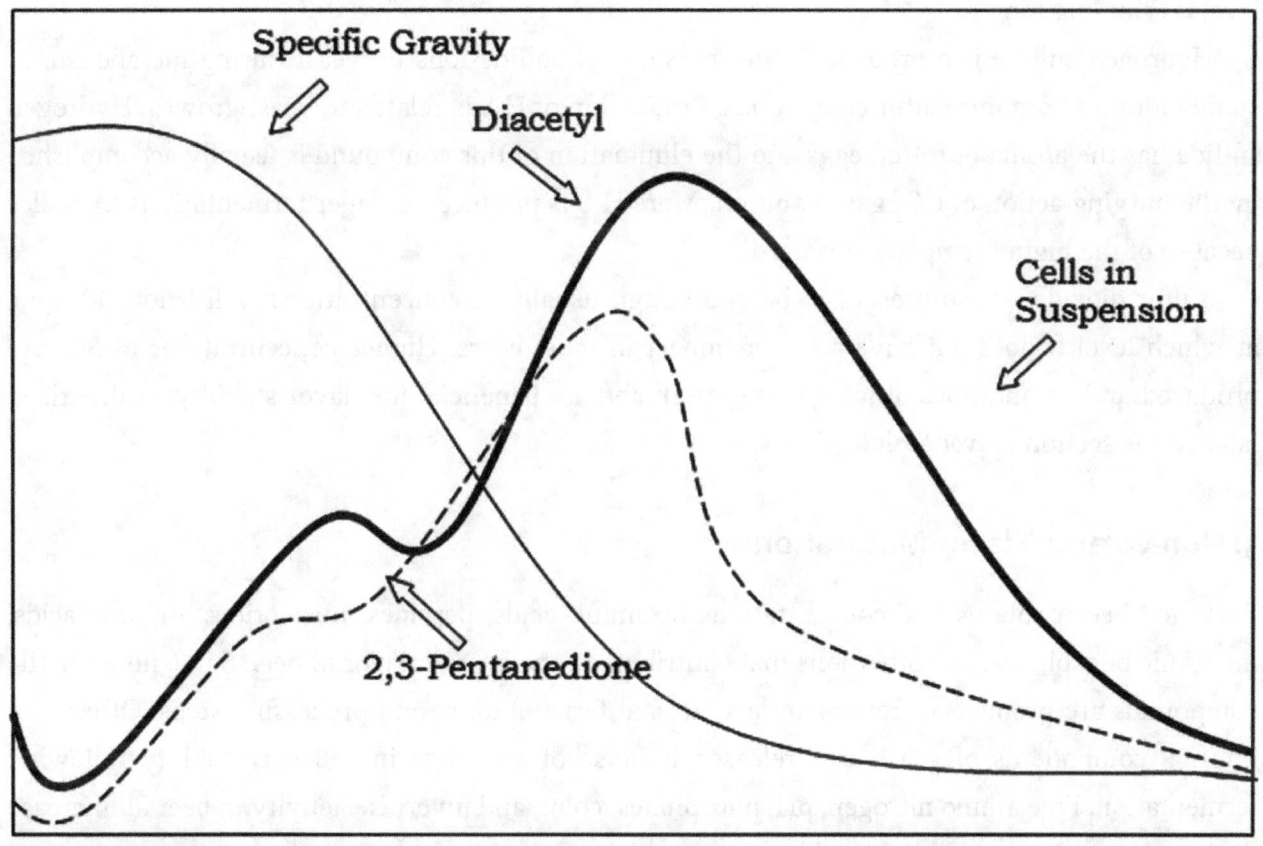

Elapsed Time (Days)

Figure 9.2: Diacetyl Profile

2. Sulfur Compounds

The subject of sulfur compounds in brewing is broad and complex. They are particularly important because of their very low flavor threshold and a flavor perception that is objectionable. Important sulfur compounds result from yeast metabolism, but many that are present in beer come from malt and hops. Yeast requires sulfur-containing compounds for synthesis of proteins. Sources of sulfur in wort for yeast metabolism are sulfate ions from water, and thiols and sulfides from raw materials, particularly sulfur-containing amino acids. Sulfur compounds in beer arise through a combination of raw material sources, processing conditions, yeast strain, metabolism and autolysis, and microbial contamination.

Three of the more important volatile compounds are hydrogen sulfide (H_2S), sulfur dioxide (SO_2), and dimethyl sulfide (DMS). Some DMS is formed during fermentation by the action of yeast on dimethyl sulfoxide produced in the kettle, although most DMS comes from the conversion of a precursor in malt during kettle boil. DMS at low concentrations is believed to make a positive contribution to beer flavor. DMS concentration levels can be determined during maturation, with the objective of achieving the desired level for the final beer. These levels, however, vary considerably for different beers. The flavor threshold of DMS is 35-40 µg/l, and each brewery should know the level it is aiming for.

Hydrogen sulfide is a product of the transport of sulfide ions by yeast during metabolism of sulfate ions and organic sulfur compounds. Production of H_2S is related to yeast growth. Hydrogen sulfide has the aroma of rotten eggs and the elimination of this compound is usually accomplished by the purging action of CO_2 gas evolution. More H_2S is produced in lager fermentations than ales because of the higher temperatures used.

Sulfur dioxide is also present in beer although usually in concentrations well below 10 ppm, at which level it does not have a flavor impact in most beers. Higher concentrations of SO_2 are produced under conditions of low yeast growth and are beneficial for flavor stability, as described later in the section *Flavor Stability*.

3. Non-volatile Flavor Maturation

Packaged beer contains low concentrations of amino acids, peptides, nucleotides, organic acids, inorganic phosphates, and other ions that contribute to the overall flavor of beer. Some non-volatile compounds are products of raw materials, normal fermentation, and processing steps. Others are internal components of yeast cells released because of a change in cell permeability following fermentation. Free amino nitrogen, pH, phosphates, color, and invertase activity in beer all increase during storage. It would be reasonable to assume that these increases are dependent on temperature, time, yeast strain, physiological condition of the yeast, fermenter geometry, and so on.

It is important to note that these changes in non-volatile compounds are not necessarily undesirable. Non-volatile compounds can contribute to palate fullness or mouth-feel, and act synergistically with other flavor-active substances and contribute to the overall flavor quality of beer.

4. Yeast Autolysis

Yeast autolysis is refers to the dissolution of dead cells by their own enzymes—the autolysis products released into the beer result in a sharp, bitter taste and a yeasty aroma. Autolysis occurs under conditions of starvation and high temperature. Holding a fermentation vessel at high temperature (> 15°C) in order to facilitate conversion of the VDK precursors is a condition that can lead to autolysis. Therefore, it is important not to allow the yeast to remain in beer for long periods at high temperature.

Lagering and Secondary Fermentation (Kräusening)

The term 'lagering' comes from the German verb *lagern*, which means 'to store', 'to age', 'to lay down'. The use of this term in the brewing industry is often synonymous with aging and storage, and sometimes other terms that are a consequence of aging such as maturation, conditioning, and secondary fermentation. The term lager beer follows from historical aging practices before refrigeration.

For clarity, the following definitions will be used:

Primary or main fermentation: The initial fermentation, during which most of the carbohydrates in the wort are assimilated. If no secondary fermentation is done, then all carbohydrates are assimilated during primary fermentation.

Secondary fermentation: Fermentation subsequent to transfer of the beer from primary fermenters with some yeast and fermentable carbohydrates present, during which residual carbohydrates are assimilated. This process is usually done at a reduced temperature. Secondary fermentation is not always used in modern practice.

Kräusening: The addition of fermenting wort to the secondary fermentation. Kräusening is not always used in modern practice.

Lagering: Non-specific term applied to aging and other processing following primary fermentation. Historically, lagering included a secondary fermentation followed by a long, cold storage period.

Ruh storage: A practice that refers to cold aging of beer with some yeast present following the completion of fermentation.

Maturation, aging, storage, and conditioning: Terms used interchangeably to refer to the maturing of beer flavor. A secondary fermentation may or may not be included.

Kräusening

Kräusen is a German term meaning 'rocky head'; and, in brewing, the word refers to the appearance of the foam head in the primary fermenter. When fermentation is most active, foam formation is greatest and the fermentation is at 'high kräusen'.

In the practice of Kräusening, beer is transferred to the storage cellar after primary fermentation, usually with some residual fermentable extract remaining. Then a volume of high kräusen beer, about five to twenty percent of the primary fermented beer volume, is added to the tank. The secondary fermentation continues as in the traditional process, except more rapidly. The degree of secondary fermentation can be controlled by the amount of residual fermentable extract at the transfer and by the amount and fermentable extract of the high kräusen added. In Kräusening, a somewhat more flocculant strain can be used because the secondary fermentation is more vigorous than without kräusen, and a flocculant strain is not needed.

Cooling may occur gradually during secondary fermentation, or rapidly at the end to promote yeast settling. The CO_2 produced helps to reach the packaged beer level of carbonation. However, the introduction of high kräusen adds more fermentable extract and produces more flavor compounds. Off-flavors such as H_2S and diacetyl are also increased after being at low concentrations at the end of primary fermentation. Lengthy storage may be required to reduce these undesirable flavor notes to acceptable levels.

Lagering without Secondary Fermentation

Historically, lagering employed shallow open fermenters for primary fermentation, but closed vessels for secondary fermentation in order to maximize carbonation. Modern equipment for refrigeration, carbonation, filtration, and so on, prevents the need for secondary fermentation and a long, cold storage period. Modern practice has shortened fermentation and lagering times, and uses rapid cooling after fermentation to aid yeast settling. If the wort is fully attenuated during primary fermentation, there is no need for secondary fermentation, and the aging process is principally for flavor maturation.

As described earlier, one of the more important classes of compounds involved in flavor maturation are VDKs. Because the rate of conversion of VDK precursors is temperature dependent, elevating the temperature can be used to hasten the conversion. Short, warm lagering has proven quite effective with minimal deleterious effects on beer quality. This lagering can be accomplished in the presence of yeast by extending the residence time in the fermenter at the upper temperature limit after the wort is fully attenuated. If there is a lack of sufficient suspended yeast, re-circulation or 'rousing' with a CO_2 purge may help.

With modern equipment, the use of separate vessels is unnecessary; and some unit operations may be combined in a uni-tank operation. For example, after a predetermined attenuation limit has been reached, yeast can be removed from the bottom cone and the beer cooled for lagering.

Periodic removal of more yeast may be beneficial during the lagering phase to prevent off-flavor development.

There are compelling economic advantages for combining fermentation and aging in one tank. Other advantages of this concept are:

- Fewer microbiological and foam retention problems because of fewer transfers
- More efficient yeast collection
- Better control of CO_2 levels, with the possibility of eliminating carbonation
- Better opportunities for automation

Addition of Modified Hop Extracts

If modified hop extracts are used wholly or partially to replace the kettle addition of hops, they can be added to beer on transfer to aging. These extracts can be pre-isomerized hop extracts, which contain iso-a-acids (isohumulone and its homologs), or reduced hop extracts. The advantages for adding modified hop extracts to fermented beer compared to adding hops to the kettle include:

- Better utilization (efficiency) of hop-bittering acids (iso-a-acids added); more iso-a-acids surviving in the beer because loss during fermentation is eliminated
- Kettle-hopping losses are eliminated, producing product with more uniform international bitterness units (IBUs)
- Degradation of iso-a-acids during kettle boil is eliminated
- Better control of IBU in beers. IBU measurements can be done during aging, and additional extract added if the IBU is below specification

Beer Recovery

Economics

During normal production, beer is lost in the yeast, in spent filter aids, and in tank bottoms. Yeast cropped by skimming or in connection with beer decantation will have low solids in the slurry. The beer in such slurry may be over fifty percent in weight. It is estimated that up to two percent of total beer output is held up in collected yeast. Tank bottoms may be two to seven percent solids. The recovery of beer from these sources may be economically advantageous. The recovery of beer also reduces biological oxygen demand (BOD) and chemical oxygen demand (COD) in brewery effluent, thereby reducing sewer charges, an additional cost saving.

Beer can be recovered from these various sources in several ways. In some cases, yeast strain differences may play a role in the selection of suitable equipment. Methods include centrifuges, membrane or diaphragm filter presses, and other types of filters.

Quality of Recovered Beer

The microbiological stability of the recovered beer is extremely important. Proper equipment, piping, and so on, must be chosen so that acceptable cleanliness can be maintained. Minimum residence time for feedstock and recovered filtrates is essential for microbiological stability. A flash pasteurization step for the recovered beer is sometimes necessary to obtain a quality filtrate for blending.

A second factor pertinent to the quality of recovered beer is its dissolved oxygen (DO) content. Processing should be carried out under conditions as anaerobic as possible. However, it is nearly impossible to make transfers without some air pick-up. The introduction of any oxygen will contribute to flavor deterioration. Blending recovered beer at low percentages will help minimize any adverse effects of oxidation.

A third, major consideration is the clarity of the recovered beer. The requirement for clarity after recovery depends upon subsequent processing and blending. If the recovered beer is added to primary production beer during transfer to aging, further clarification occurs downstream. If the recovered beer is added later in the process, it may be necessary to filter it before pasteurization and blending.

Other properties of recovered beer that are likely to vary are the color, pH, and flavor. Color and pH changes generally are not significant because of subsequent blending. Particular attention should be paid to the flavor of any recovered beer. Flavor changes may be caused by yeast autolysis, or the recovery processing steps. Keeping the temperature of the feedstock below 5°C will help minimize flavor changes.

The recovered beer can be blended into normal production beer at any convenient step in the operation. However, the actual choice may depend on the configuration of fermentation and aging equipment, the number and types of beer being brewed, and so on. The final extract of the recovered beer may dictate the point of blending; and the recovered beer will have low carbonation. In any case, the brewer must determine a maximum percentage of recovered beer to blend into production beer. Normal practice is to use not more than ten percent. Taste testing of blended beers gives more confidence in the use of recovered beer.

Clarification

On the completion of aging, the beer contains some yeast, colloidal particles of protein-polyphenol complexes, and other insoluble material that was driven out of solution by the low pH and the cold temperature during aging. If a brilliant, clear beer is desired, the clarification must remove these substances before beer packaging can be done. Four basic clarification techniques are used either separately or in combination:

- Sedimentation
- Use of finings
- Centrifugation
- Filtration.

Gravity Sedimentation

This is surely the simplest method for achieving clarity and was the only method before the development of centrifuges and filters. Historically, the chilling of fermented beer to about 0°C for long periods promoted the sedimentation of yeast and other particles. However, despite its simplicity, caution is needed because yeast autolysis occurs readily, especially if the packed yeast mass begins to heat. With clarification by sedimentation, beer losses are relatively large, and cleaning of the tank bottoms is costly.

Finings

Although good clarity can be obtained from simple sedimentation, better results can be obtained in less time by using fining agents. Because of their chemical structure, they carry a net positive charge and interact with yeast cells, which are negatively charged, and with negatively charged proteins. Negatively charged proteins have been implicated in haze formation. Consequently, removal of these compounds improves physical stability. Finings increase the volume of tank bottoms and also increase tank clean-up costs and beer losses. The most common fining agent is isinglass, which is made by chemically treating the swim bladders of certain fishes and types of clay. The use of finings improves subsequent filtration.

Filtration

Filtration generally refers to clarification of beer through several stages to produce a crystal-clear product. The purpose is to remove suspended material and residual yeast, which would otherwise cause the beer to be hazy. The particle size of suspended material in beer is 0.5–4 μm. Particle size information is necessary for the brewer to set filtration parameters.

The mechanisms of filtration can be classified into three types:

1. Surface filtration
2. Depth filtration through mechanical entrapment of particles
3. Depth filtration through adsorption of particles

Surface filtration means that particles are blocked at the surface of the filtration medium because the particles are larger than the pores in the medium. In depth filtration, particles pass into the filtration matrix; the particles are either mechanically trapped in the pores or adsorbed on the surface of the internal pores of the filtration medium.

Filtration may be used at two or more stages after aging, depending on the particulars of cellar operations. The terminology for various filtrations in cellar operations differs from brewery to brewery. The first or primary filtration stage removes the bulk of yeast and suspended material and the second stage produces a brilliantly clear beer. The addition of stabilization agents occurs before primary filtration, and these are substantially removed by the filter. Primary filters are almost always powder filters. A turbidity sensor can be installed at the outlet of the filter to monitor filter performance.

As a second stage, polish or final filtration removes any additional suspended solids resulting from lagering at cold temperatures and any adsorbents added for stabilization. These final finishing steps are generally preceded by a final beer cooling operation to aid precipitation and ensure that the beer reaches the government cellars at the proper temperature. Polish filtration may consist of two separate filters. After a first filter, trap filters may be used as an immediate final stage only to guard against any breakthrough from the upstream filter, not to perform further filtration. Trap filters are usually membrane filters. There should be no further addition of any substance to the beer stream after the last filter, as the introduction of unfiltered liquids may prove harmful to the clarity of packaged beer. Sterile filtration to remove bacteria present in the beer is described in the section *Sterile Filtration.*

Filters

The first categories of filters to be discussed use powders or filter aids and are the most popular. The materials for powder filtration include Kieselguhr (diatomaceous earth or DE) and perlite (volcanic silicate). Then, those that use sheets, cartridges, membranes, and so on, will be described. Filters are used not only to clarify beer but also to clarify wort, recover wort from separated trub, and recover beer from tank bottoms.

Diatomaceous earth is usually calcined after mining in order to eliminate organic matter. The high porosity of the diatom skeletons is ideal for filter beds, as the liquid passes through the bed while the suspended particles cannot. Diatomaceous earth is supplied in a variety of grades from which the brewer chooses to accomplish clarification objectives. The different grades have particle size distributions that affect filter flow rates, filter bed permeability, the degree of filtration (coarse to fine), and so on.

Perlite is an ore of volcanic rock containing silica. When crushed and heated, perlite expands to become a light, fluffy powder and is suitable as a filter aid. The expanded perlite is milled and graded, producing filter aids with a range of permeabilities. For any filter aid, the important properties are: good permeability to keep the pressure drop low across the filter and good wetting to ensure uniform dispersion and bed formation.

Filters that use powders are sometimes called DE or Kieselguhr filters. Filters that use filter aids (powders) operate on a principle of building a bed or cake of powder on a septum or filter screen. The porous bed creates a surface that traps suspended solids, thus removing them from the beer. Normally, the filter septum is pre-coated with a filter aid in advance of the beer filtration run. This pre-coat forms the base layer for the bed. The rough beer to be filtered is dosed with more filter aid, called body feed, at a concentration based on the solids content to be removed. The use of body feed helps to achieve the goal of maximum filter throughput. As beer is run through the filter, the bed increases in thickness because of the body feed, thereby maintaining bed permeability. Various grades of powders are used, depending on the filtration performance desired and beer to be filtered. For example, primary and polish filtrations will use a different grade and thickness of pre-coat. Body feed may not be required in polish filtration as it is in primary filtration. The different types of DE filters that follow are simply different implementations of these principles.

Filters are operated until the differential pressure rises beyond a designated point, which requires the flow rate to be reduced, or to the point when the bed depth reaches a thickness that bridges the spaces between the septa in the filter. Filter systems are designed to function within a specific range of pressures. An excessive differential pressure can cause: (a) the filter leaves to collapse, (b) the filter shell to burst, or (c) the pumps to fail, as they are not sized to operate with the increased energy needed to maintain the flow rate.

The relationship between filter operational parameters is:

$$\Delta P = \mu V \times L/\beta$$

Where,

   ΔP = Differential pressure

   V = Specific flow rate (flow rate of beer per unit filter area)

   μ = Beer viscosity

   L = Bed depth

   β = Bed permeability

The specific flow rate will depend on the differential pressure or pressure drop across the filter bed. The differential pressure is directly proportional to the specific flow rate, beer viscosity, and bed depth, and is inversely proportional to the bed permeability. Therefore, at a constant flow rate and viscosity, filter performance depends on the ratio of the bed depth to its permeability. Diatomaceous earth is added to the beer (body feed) being filtered in an effort to maintain permeability as filtered particles from the beer reduce the bed permeability. However, this ratio inevitably increases gradually, causing the differential pressure to rise to an unacceptable level resulting in the end of the filter run.

Plate and frame filters are traditional models that are still used in the industry. They consist of a series of parallel plates covered with filter sheets used to support the filter bed. The frames between the plates control the bed depth. Different numbers of plates can be used, depending on requirements. Most filters allow beer to pass through both sides of the plates, thus doubling the surface area per plate.

Leaf filters consist of a series of circular, stainless steel leaves as perforated support plates. The leaf configuration can be horizontal or vertical. The leaves in horizontal filters have a stainless steel woven septum to support the bed, while vertical filters use the septum on both sides. Their operation is quite similar to plate and frame filters.

The candle filter is of a different design entirely, although the filtration principle is the same. The candles can be porous ceramic but are usually perforated or fluted, stainless steel tubes covered or surrounded by a stainless steel support of various types. This rigid septum is easier to clean than filter leaves used in the powder filters. There is also an operational advantage. The beer is fed to the outside of the candles and the filtrate is collected through the inside. The circular design means that the increase in bed thickness during operation is less than other filters, and the pressure drop increase occurs at a slower rate. The ceramic filter can be used for sterile filtration of beer.

Sheet filters are similar in design to plate and frame filters. Whereas the sheet used in powder filters acts as a septum to hold the pre-coat, the sheet acts as the filtration medium in the sheet filter. The sheet is usually made of cellulose impregnated with DE. Other materials can be added to achieve both the desired liquid permeability and solids retention. These filters have wide applicability, but are generally used after a primary DE filter because they do not have the capacity of the powder filters. They are also suitable for sterile filtration. Most filters of this type can be easily backwashed and several runs can be made before replacing the sheets. Two disadvantages are the high labor cost of handling the sheets and lack of automation. Sheet filters are often used for keg beer.

A related type of filter is the pulp filter. The cellulose and cotton fibers are formed into circular pads and joined face to face. The filter can be used for primary filtration or later filtration stages and is suitable for sterile filtration. The pads can be reused by washing the material after dispersing the pad fibers in water. Because the pads are reusable, disposal problems are reduced considerably. The use of these filters is very labor intensive.

Cartridge or membrane filters are generally much smaller and serve as sterile filters, and as trap filters that catch breakthroughs of DE occurring upstream. The filter medium is usually a membrane produced from polymeric synthetic materials, such as Nylon 66 or cellulose esters. With man-made materials, the membrane can be constructed to a desired permeability and mechanical strength with a large surface area. These systems are generally economical and easy to maintain.

Sterile Filtration

Sterile beer filtration is defined as an operation that produces sterile beer ready for packaging with no subsequent pasteurization. As discussed previously, several filter types are suitable for this task: sheet, membrane, ceramic candle, and pulp filters. The type of filter selected for sterile filtration will depend on the brewer's needs and on appropriate features, such as throughput, ease of maintenance, cleaning, and sterilization.

Whichever type of filter may be used, it must be preceded by at least one other filter that can remove the entire colloidal load, including chill proofing agents, and reduce the yeast count, preferably near zero. Typically, the bulk of suspended particles are removed with a DE filter, followed by a DE, sheet, or cartridge filter that will remove residual material sufficiently to reach a haze specification prior to the sterile filter. The sterile filter acts as the final trap of yeast and bacteria.

Sterile filters are not absolute filters. Therefore, the brewer will set a specification for the maximum concentration of bacteria in sterile-filtered beer. As it is possible for a single beer-spoiling bacterium in a bottle or can to spoil the beer, there is a need to balance the risk of spoilage against filter practicality and throughput.

Suitable microbiological sampling and methodology is needed to measure adherence to specifications. Rapid microbiological assay methods are of particular importance to reduce the quantity of product awaiting release to packaging.

To determine if the filtration system will allow microbiological specifications to be met, the brewer must measure the efficiency of the system for removal of microorganisms from beer. This is also called challenge or integrity testing. For example, for a specific yeast/bacterial load in the beer, there is a measured reduction in that load in the beer filtrate. Beer or water is seeded at a known concentration with a beer-spoiling bacterium. Under fixed filtration conditions, beer filtrates are collected and plated. The ratio of colony-forming units before filtration to after filtration, that is, the change in microbiological load across the filter, is called the log reduction value and is expressed as a logarithm. For example, a ratio of 1×10^9 means the system has a log reduction value of nine. A log reduction value of eight to nine is required of a filter system if it is to be useful as a sterile filtration system. Filter media for sterile filtration, particularly sheets and membranes, will have specific log reduction values that help the brewer optimize the system. Filters must be tested with appropriate bacteria. The brewer should select beer-spoiling bacteria common to the brewery to obtain a practical measure of their filter integrity.

Based on the filter system log reduction value and the filtered beer specification, the incoming beer may present a greater microbiological load than that for which the filter is designed. In such cases, sanitation procedures further upstream in the process must be addressed. The sterile filter cannot be expected to remedy poor microbiological practices upstream. In the end, instilling a proper attitude toward sanitation and care with regard to producing sterile beer is invaluable to reducing microbiological problems.

Transfer to Packaging

In the finishing steps of the transfer of aged beer to packaging, a major concern is oxygen pick-up. Chill proofing, dilution, carbonation, and final filtration steps generally occur in a continuous sequence leading to package release or government tanks. These operations are separated by transfer tanks and connected by pipes. Minimizing oxygen pick-up in these vessels and piping/pumping systems is important because it is difficult to correct a package release tank that has a high concentration of dissolved oxygen.

To minimize oxygen pick-up, filter feed, surge, and transfer tanks should be purged with CO_2 or packed with carbonated water prior to use. For filtration, oxygen in filter pre-coat and body feed in makeup tanks is reduced by CO_2 purging for a sufficient time. Using deaerated water for makeup is also helpful.

A critical area to reduce oxygen pick-up is in the filters themselves; they are usually opened to the atmosphere for cleaning. Even when a closed filter is purged free of the filtration medium, sluicing with water that has not been deaerated presents a risk; the filter may require CO_2 purging. Deaerated water can also be used to purge transfer lines. Account must be made for beer dilution from water left in filters, tanks, and transfer lines.

If packaged beer will not be pasteurized, the transfer of sterile beer to packaging presents additional challenges because the contact of sterile beer with any surface presents an opportunity for contamination. Generally, it is advantageous to dedicate specific transfer and package release tanks for sterile-filtered beer. This reduces the possibility of contamination of tanks and transfer lines from beer that is not sterile. The number of dedicated tanks must be chosen to buffer the sterile filter output with packaging requirements. To prevent contamination of sterile beer further downstream, the use of dedicated bottle and can packaging lines is advantageous.

Beer Stabilization

The stabilization of beers may refer to flavor stability or microbiological stability, although stabilization commonly refers to physical characteristics.

Flavor Stability

Chemical reactions continue to occur after the beer is packaged. Many of the changes that lead to stale beer flavor are caused by chemical oxidation. Flavor stabilization, then, generally refers to the protection of beer from oxidative changes. In early research, the cardboard flavor of stale beer was attributed to trans-2-nonenal. Furfural and related compounds have also been identified in staled beer. Stale off-flavors are generally attributed to the oxidation of higher alcohols to aldehydes by melanoidins, but there are many more chemical routes participating in the staling of beer. The topic of flavor stability is far ranging and complex, and only two important factors, SO_2 and oxygen, that have a clearly established effect on flavor stability.

Sulfur dioxide, in the form of bisulfite ion, protects against oxidative flavor in two ways. It reacts with oxygen, eliminating it from beer and potential oxidation of beer components. It also reacts with aldehydes, which have stale flavors rendering them flavor inactive. While the complex of bisulfate and unsaturated aldehydes is irreversible, the complex with saturated aldehydes is reversible as other chemical species compete for the bisulfite. Sulfur dioxide in beer occurs naturally from fermentation and is increased under conditions of low yeast growth. In addition to naturally occurring SO_2, one can add antioxidants to beer. Potassium meta bisulfite or forms of ascorbic acid are sometimes added after fermentation as reducing agents to counteract oxidative changes.

Excluding oxygen from beer is an important step in flavor stability. Because yeast is an oxygen scavenger, once it is removed, any oxygen picked up in processing has the potential to oxidize beer. Therefore, flavor stability is enhanced by excluding oxygen from the beer during aging and finishing operations after the yeast is removed. The use of CO_2 to pack tanks and to transfer beer reduces the possibility of air pick-up. Flavor stability may be enhanced by proper handling of wort in the brewhouse, by reduction of oxygen pick-up during mashing, lautering, and wort cooling, and so on.

Biological Stability

Microorganisms can contribute to flavor instability. Certain bacteria (for example, *Lactobacillus sp. and Pediococcus sp.*) and yeasts of the *Saccharomyces* and *Hansenula* genera can spoil beer by producing undesirable flavor compounds, such as VDKs and lactic acid. Generally, brewers conduct microbiological tests specifically for beer spoilage microorganisms. Microorganisms can also grow and form a haze by increasing their number. Proper pasteurization ensures biological stability but requires the heating of beer, which accelerates potential oxidative flavor changes. Biological stability can be achieved by sterile filtration in which microorganisms are removed by special filtration systems. Although sterile beer can be produced by available filtration technology, contamination is still possible during filling and keg operations. In fact, aseptic filling is more difficult than producing sterile beer by filtration.

Physical Stability

Colloidal or non-microbiological haze is a result of the precipitation of insoluble complexes formed from beer constituents. The general components are known, but the mechanisms of interaction and complexation are not well understood. There are two types of physical haze—chill haze, which appears when the beer is chilled but dissolves upon warming; and permanent haze, which never fully dissolves under any condition. Beer affected with permanent haze remains cloudy and may even develop sediment.

Research has shown that several chemical species are present in haze material. The major component appears to be proteinaceous material in the range of 1000–40,000 Da. Other observed components of colloidal haze are polyphenols, and to a lesser extent metal ions and polysaccharides. It is generally believed that complex proteinaceous compounds and polyphenols become associated through hydrophobic and hydrogen bonds involving proline residues of proteinaceous compounds. The presence of oxygen may play a role in polymerizing the phenolic constituents. Those portions of the proteins and polyphenols that contribute to haze formation are referred to as the haze-active fractions.

Knowing that haze consists of insoluble protein–polyphenolic complexes, preventative measures can be directed at one or both of these classes of soluble compounds. Three methods are used to 'chill-proof' beer, as physical stabilization is commonly called—treatment with proteolytic enzymes, use of finings such as tannic acid, and adsorption.

Brewhouse Procedures and Filtration

Some remedial measures can be taken during brewing to improve physical stability and to reduce the need for stabilization. However, removal of proteins and polyphenols must be done carefully as both contribute to the character of beer—both its flavor and physical characteristics.

Selection of malt with lower soluble nitrogen, good modification, and high diastatic power can lower the proteinaceous content of beer. Proteolysis in the mash tends to reduce protein content, although this reduction of high molecular weight protein in the mash may be due to a precipitation mechanism rather than an enzymic one. When adjuncts are used, especially corn or rice, the wort protein level is reduced proportionally. Lower mash pH reduces the solubility of polyphenols. The last running from sparging can be high in polyphenol content if the sparge water pH is not carefully controlled.

Proper adjustment of wort boiling helps control the levels of polypeptides and polyphenols. A long and vigorous boil helps coagulate the complexes, and the presence of oxygen will aid the oxidation of polyphenols. However, oxidation of compounds important for flavor stability may also occur. A good, hot break along with efficient wort clarification enhances physical stability.

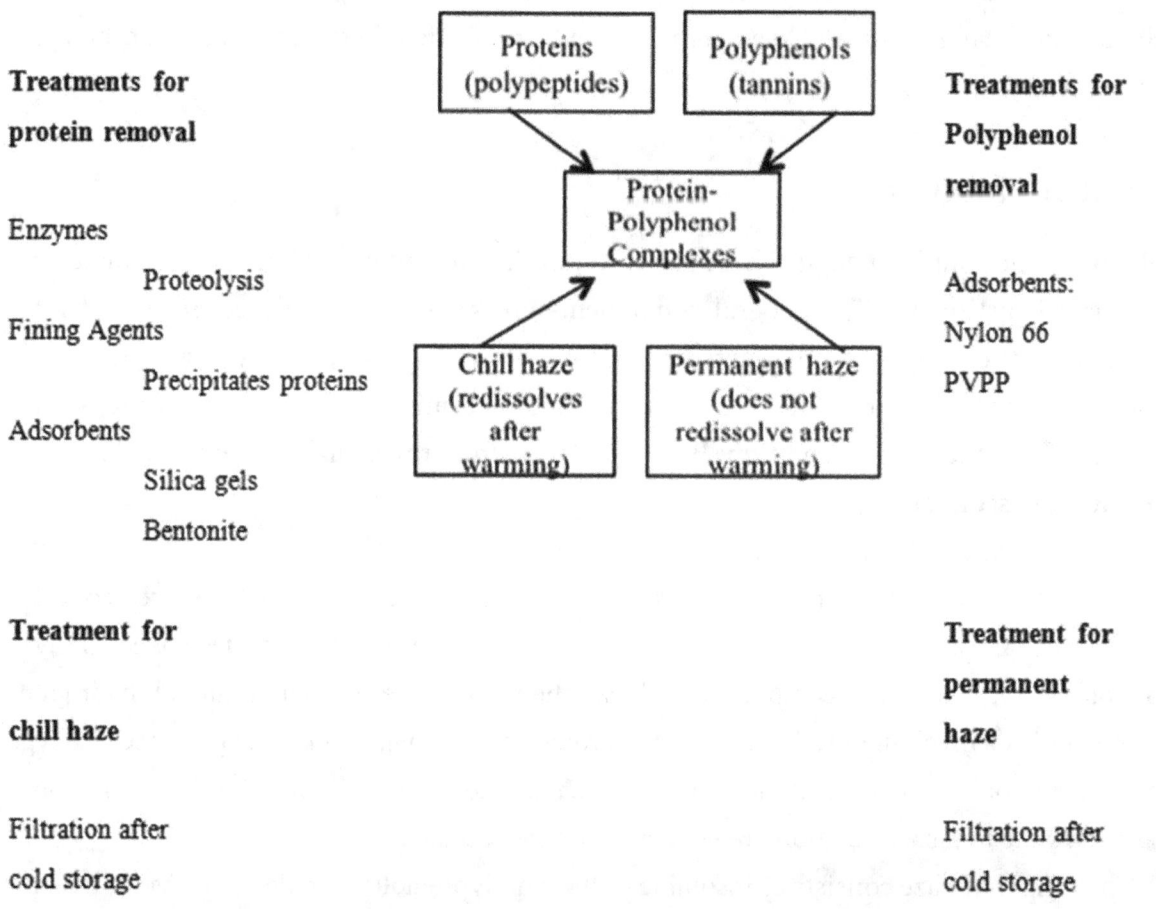

Figure 9.3: Chill Proofing Strategies.

Measurement of Haze

The measurement of haze or turbidity is based on the principle of nephelometry in which light reflected from particles in solution is measured. The angle of reflection is usually ninety degrees, although smaller forward scattering angles are also useful. The measurement of turbidity depends on the color of the incident light and on the size and shape of the light-scattering particles. Calibration of instruments specifically designed for nephelometry depends greatly on stable particle standards. Formazin is usually used, but more expensive, chemically polymerized spheres can be used. Use of instruments is further complicated by imperfections in the measuring cells. Beer haze determined in bottles introduces large, random errors, whereas the use of optical cells is time consuming. Another complication arises because different instruments produce different results on the same samples, and differ in their responses to particles of different sizes. In-line turbidity meters are often difficult to correlate with laboratory instruments.

It is also possible to rate beer haze visually by comparing the sample with standards usually based on different concentrations of formazin (ASBC Methods of Analysis: Beer-27, A). With this method, there is difficulty in obtaining agreement between individuals on the level of turbidity in a sample. At best, this method is qualitative.

Another element of confusion is that different measurement units are used; for example, the American Society of Brewing Chemists (ASBC) and the European Brewery Convention (EBC) use different units of measurement. The major problem, however, lies in trying to quantify the human perception of hazy beer by quantifying particles with a range of sizes and shapes and other light-scattering characteristics; the correlation is not always good.

Carbonation

Basics of Beer Carbonation

Carbon dioxide solubility in beer is usually measured in volumes of CO_2 per volume of beer at standard temperature and pressure. This means that one volume of CO_2 is equal to 0.196 percent CO_2 by weight or 0.4 kg CO_2/hl. Typical American lagers contain 2.5–2.8 volumes of CO_2. Ales are generally lower. Because beer contains 1.2–1.7 volumes of CO_2 after a normal non-pressurized fermentation, another 1 volume or about 0.5 kg/hl must be added before packaging. Considering that other uses for CO_2 in the process consume CO_2, it is generally economical to recover excess CO_2 from fermentation. In some breweries, losses together with requirements may exceed recovery and CO_2 must be purchased, although under careful conditions breweries can be self-sufficient. The amount of CO_2 in the solution depends on Henry's law. This law states that the amount of a gas dissolved in a liquid is proportional to the concentration of the gas in the headspace. Therefore, CO_2 concentration in the beer can be increased by increasing the head pressure of CO_2. Temperature changes the solubility. A temperature increase leads to a decrease in solubility. Therefore, the desired CO_2 concentration can be attained by fixing the temperature and pressure at appropriate settings.

Because the solubilities of gases are independent of each other (Henry's law), the level of carbonation has no influence on oxygen pick-up as the product moves through the process.

The time required to reach a desired CO2 concentration depends on physical factors. Finer bubbles have more surface area per unit weight and dissolve faster than larger bubbles. Moreover, finer bubbles rise more slowly. The longer it takes for bubbles to rise through a tank, the more time there is for solution. Therefore, carbonation stones are designed to form a fine mist of bubbles. If the headspace is filled with CO_2, a larger headspace–liquid interface area will shorten carbonation time. The solution of CO_2 also slows as equilibration is approached.

Pressure and temperature relationships to CO_2 concentrations are used to establish a tank concentration. Measurement of CO_2 in tanks can be done with a sensor separated from the liquid by a membrane permeable to gases. A common alternative to sensors is the Zahm–Hartung method. A metal bottle is filled under controlled temperature and pressure. After establishing equilibrium with the headspace, the temperature and pressure are read and converted by means of a table (from *Methods of Analysis*: Beer-13, A; ASBC) to volumes of CO_2. Corrections can be made for oxygen and nitrogen to improve accuracy. In tall tanks, the CO_2 concentration will be higher at the bottom because of the greater hydrostatic head.

Modern Carbonation

Carbonation can now be done by in-line injection or by in tank carbonation. In-line injection can be done whenever beer is transferred. However, it cannot be done upstream from DE filtration because CO_2 bubbles would disturb the filter bed. In tank carbonation usually involves introduction of CO2 through a carbonation stone in the bottom of the tank. The purpose of the stone is to form fine bubbles of CO_2, which readily dissolve in the beer. Another reason for carbonating in tanks is that oxygen and objectionable aromas can be swept out of the beer if the tank can be open to the atmosphere during the early part of the process. The tank is then closed and carbonation begins.

Standardization

Blending for Consistency

Blending or standardization refers to the mixing of different batches of beer to achieve product uniformity. Generally, blending is done to achieve an exact alcohol concentration, specific gravity, or original gravity. Blending can be done to achieve uniformity in other parameters, for example, bitterness units or color.

Occasionally, blending is used to attenuate an objectionable flavor note. For example, if a fermentation problem led to a high diacetyl concentration or noticeable sulfury character, the beer could be blended with a normal product in an attempt to dilute the objectionable flavor. Blending guidelines are established by brewers to prevent noticeable deviations from flavor uniformity.

10

Packaging

Introduction

"Packaging means all products made of any materials of any nature to be used for the containment, protection, handling, delivery, and presentation of goods."

– *Handbook of Brewing*, Ed. by Hans Michael Eblinger

Beer has some additional requirements when it is packaged, which do not necessarily apply to all types of food and drink.

The first is pressure—a beer package needs to be able to withstand the pressures generated when a carbonated liquid is heated, either during pasteurization or when transported at high ambient temperatures.

The second is light—beer is degraded by sunlight, and this must be avoided. It is surprising then that glass is so often used for packaging. By using colored glass together with additional external protection and covered transport, this problem can be largely avoided.

The final factor is impermeability to gases—it is vital that the gas composition of beer is not altered in its package. This means not only preventing CO_2 from getting out, but also preventing access to oxygen.

Packaging is the most expensive aspect of brewing, representing up to two thirds of the cost of beer production. It is also one of the least forgiving steps in the brewing process.

A perfect batch of beer can be ruined in nanoseconds by microorganisms on the nozzles or too much oxygen making its way into the bottles or cans. Foreign objects like glass could also get into the finished product. Too little glue on the corrugated boxes can send the product (and profit) crashing to the ground.

A properly designed package must enhance the value of its contained product and impact an impression, directly or delicately, on the consumer—understanding this is paramount to understanding packaging. *Packing is not equivalent to packaging. The 'package' is used primarily to move goods through a distribution system to the consumers. On the other hand, packaging is the term used by the marketing for the consumer—'Buy our product, please'. Packing is, simply, a tool of packaging.* (*Handbook of Brewing*, Ed. by Hans Michael Eblinger)

Packaging must accomplish four goals:

1. Contain and protect the product
2. Convenience of use for the consumers
3. Communicate the contents
4. Motivate consumers to purchase the product

All of these must be balanced to provide the most desirable result.

Packaging Materials

The materials used for packaging beer are widely used elsewhere, but there are certain nuances in the way beer is packaged that make it distinct. None of the primary materials used have ideal properties, but our packages are designed to make the best of what we have.

Glass has several drawbacks. It does not have an easily printable surface, so we use labels. It is not opaque, so colored glass is generally used to protect the beer. It is also heavy and breakable. However, on the plus side, it is inert, attractive, reusable, and well established.

Steel and aluminum fit the ideal more closely, but they have to be coated in various ways to make them inert and to facilitate printing and decorating.

Plastics have had limited success for beer, although polyethylene terephthalate (PET) has been tried; PET works for soft drinks but it cannot be pasteurized as a whole package. Its barrier properties are not good enough for a sensitive product such as beer. This is an area where research is ongoing, and it may eventually change the market.

Wood is the traditional material for casks but due to difficulties with cleaning, it has given way to metal containers that are lighter, stronger, and easy to clean.

For secondary packaging, there is nothing unique to the brewing industry—although there are some distinctive shapes for palletizing for kegs and casks. The trend in materials for secondary packaging has been to move away from wood and corrugated board. Plastic sheeting and paperboard have gained prominence. These materials can enhance the appearance of a pack and assist in marketing. They also improve the strength-to-weight ratio, particularly for plastic shrink wrap, and this has helped in producing more durable and lighter packs and usually at a reduced cost.

Packaging and the Brewing Industry

Packaging is of enormous importance to the brewing industry for a variety of reasons, and these will probably increase with time. Brewing has moved, over the last century, from being a small-scale industry where the product was consumed from glass bottles, cans or kegs, to a large-scale industry where beer is transported across huge distances. The product is expected to be uniform with a long shelf life rather than one that is instantly consumed (before it can deteriorate). Therefore, it is vital that beer quality be maintained, and this has become a huge technical challenge.

Glass Bottles and Bottling

Glass

Glass has been used as a packaging material for centuries, and its use for beer goes back a long way. Its use in the brewing industry increased rapidly with the mechanization of the glass molding process about a century back. Bottles for beer are a major part of the glass market in most countries.

The three main colors for beer bottles are clear (white flint), green and amber. Color is important from the viewpoint of quality.

Colorless (White Flint) Glass

To obtain completely clear glass, it is necessary to have raw materials with no impurities. The main impurity is usually iron oxide and its levels need to be below 0.04 percent—otherwise, the glass will have a blue-green tinge. To a certain extent this can be neutralized by adding 'decolorizers', which are traces of cobalt and selenium.

Green Glass

Green glass is obtained by adding small amounts of iron oxide and chromium oxide to the melted glass. Iron oxide on its own gives a pale green color at levels of about 0.15 percent.

Amber Glass

Amber is probably the most common color for beer bottles and is obtained by adding carbon as a reducing agent to melted glass with moderate levels of iron oxide. It also requires a trace of sulfur.

Crowns

Metal crowns for bottles have dominated the bottle top—the outer edge has twenty-one serrated 'teeth' and the underside is coated with a flowed-in plastisol liner.

Nowadays, crowns are manufactured by the billion all over the world and are regarded as the standard closure for beer and soft drinks bottles. They have the benefit of low cost, reliability, convenience, and pressure resistance. In addition to the universal prize-off crown, there is also a significant market for twist-off crowns. Crowns are made of low-carbon steel or sometimes stainless steel, which is supplied as sheets with a thickness of (usually) 0.24 mm, although other sizes are also used.

Adhesives

The majority of labels in use are still paper-based, and wet adhesives are used to stick them to bottles. These adhesives fall into two main categories—caseins and resins.

Casein is a protein that dissolves in dilute aqueous ammonia; when various stabilizers and additives are added, it produces a powerful adhesive for label application. These casein-based adhesives have the advantage of being removed by caustic in a bottle washer. They are fast setting and even work below freezing point.

Resin-based adhesives are derived from starches and dextrins and are also soluble in dilute aqueous ammonia. Self-adhesive labels have grown in popularity in recent years. These labels are mounted on a smooth backing strip and peeled off as they are applied to the bottle. They have the advantage of instant adhesion to the bottle and can be more precisely positioned. They tend to be used with non-returnable bottles. A problem with returnable bottles is label removal, and this is hindered if the adhesive or paper is resistant to wetting as a result of paper coating or lamination.

Bottling Plant

Bottling machinery has evolved over a considerable number of years but continues to evolve, as the demand for increased speed, quality, and level control increases. Progressive improvements in automation have meant that numbers of operators have steadily fallen from dozens to a mere handful. The stages in the operation of a bottling line are as follows:

- Offload empties
- Wash/rinse empties
- Inspect empties
- Fill and crown
- Pasteurize
- Contents check
- Label and inspect
- Cartons packer
- Palletize

There are two main types of plants—returnable lines with a bottle washer, and non-returnable lines with a rinser only. In practice, however, lines often have both facilities.

Bottle Rinsing

New non-returnable bottles have very little in the way of contaminants, but in order to ensure cleanliness they are normally rinsed before filling. Rinsers come generally in two forms—a rotary machine where the bottles are gripped at the neck and then inverted for spraying, and linear machines with an inverted belt where the bottles are gripped at the neck, inverted and rinsed, and then placed back on a conveyor.

Bottle Washing

The washing of beer bottles on a returnable line is a major component of the operation. There are a number of variables that can affect the operation:

- Temperature
- Detergent strength and composition
- Bottle condition
- Water quality
- Contact time

Bottle washing machines are designed to clean the bottles by using a combination of steeping and jetting, so that heat, and chemical and mechanical actions are used to remove labels, glue, foil, dirt, and residual beer. The combination of soaking and jetting is the norm in today's machines. Bottle washers are large machines and work by loading up rows of dirty bottles into pockets on a continuous carrier chain, where they are held until they are discharged clean at the end of the cycle. There are two basic types of machines:

- Single-ended—where the loading and removal of bottles is done at the same end; one operator can easily monitor these machines
- Double-ended—where the discharge end is at the opposite end to loading

There are numerous varieties of internal layout for bottle washers.

The continuous belt that carries the bottles snakes around inside the machine in a set pattern so that the bottles are put through a series of cycles of jetting, inversion, draining, steeping, rinsing, and so on until they are clean.

A typical sequence of events inside the washer would probably be as follows:

1. <u>Pre-soaking and rinsing</u>: The first requirement once the bottles are loaded up is to drain out any residue and give the bottles a preliminary soak to remove any easily soluble residue. This also helps to warm up the bottles. The water from this pre-soaking will be high in effluents and should be continuously discharged and replenished.

2. <u>Immersion</u>: After rinsing, the bottles are immersed in the main detergent tank where the internal and external surfaces are thoroughly soaked. During this extended steep, the labels, glue and foil, as well as any beer residue inside the bottle, must be loosened. When bottles are lifted out to transfer from one soak bath to another, they are jetted with detergent to help remove labels and dirt.

3. <u>Rinsing</u>: The bottles are now conveyed above the soak tanks inverted and subjected to a number of internal and external rinses by jetting to get rid of the detergent residues and any remaining solids. After rinsing, the bottles are drained and then discharged onto the out feed conveyor to the empty bottle inspector.

4. <u>Temperature:</u> The typical temperature ranges from 60–85°C for the main detergent tank, but the temperature is ramped up gradually to avoid thermal shock to the bottles. This is where multiple steep tanks help by having individual temperature control. The presoak tank is usually at 35–40°C and the next at 55–60°C with the main soak at 75–85°C. Rinsing is carried out in stages with a drop of 10–15°C between each stage so that the bottles emerge at ambient. A temperature difference of less than 25°C between sections should be enough to avoid breakages in the machine.

5. <u>Detergent:</u> The detergent of choice for bottle cleaning is caustic soda (NaOH). A typical strength is one to three percent. Higher strengths should be avoided—otherwise, bottle etching and scuffing may increase. A number of formulated bottle washing additives are available, which help to emulsify and disperse the dirt as well as sequester metal ions and suppress foaming. Polyphosphates, EDTA, glucanases, and glucoheptanoates are common additives.

6. <u>Contact time:</u> The time needed to soak and clean bottles is temperature and caustic dependent, but a typical total cycle time is ten to fifteen minutes. There are two other aspects to bottle washing that deserve mentioning—label removal and ventilation. Labels come off in the main caustic steep tank, where the detergent is circulated to maintain its temperature and strength. The suspended labels are removed on a sieve at the infeed to the re-circulation pump and then put through a hydraulic press to remove excess caustic. Labels need to be taken out of the soak tank quickly—otherwise they will disintegrate. Ventilation of the bottle washer is needed to remove hydrogen gas, otherwise there is an explosion risk. Hydrogen is generated by the dissolution of aluminum in the foils around the bottles' necks.

7. <u>Empty bottle inspection (EBI):</u> The modern generation of EBIs use solid-state charge-coupled device or CCD optical inspection systems (i.e., cameras for checking bottles from as many angles as possible and compares the images with preset values). Any non-conforming bottles are rejected and can be manually inspected when suitable. The machines are relatively compact and do not require manual attention unless rejection rates are serious. Good bottles should have rejection levels under one percent. Faults typically seen in imperfect bottles are:
   - Bottles with residual internal dirt, such as dead insects
   - External contaminants such as traces of paper or adhesives
   - Bottles containing residual caustic
   - Defective bottle openings such as a chipped neck or damaged thread
   - Cracked sidewalls or inclusions such as gas bubbles or ceramics
   - Badly scuffed bottles that have been used too often

Bottle Filling

Large-scale bottle fillers are always rotary machines and the size is related simply to the desired capacity. The largest machines can have about 200 filling heads, be around 5 m in diameter, and produce up to 100,000 bottles per hour. Various ways of filling bottles are used, but in brewing it is always by using gas counter pressure to keep CO_2 in solution and also by the iso-barometric method— that is, the bottle pressure is the same as the counter pressure on the beer supply, so the beer runs into the bottle effectively. The bottles' contents are controlled by filling them to a predetermined height, but, recently, filling by volume alone has become available. Because of its carbonation, beer is always filled cold between 0–3˚C.

The main process objectives during filling are:

- No product loss
- Consistent contents
- No microbiological (or chemical) contamination
- No loss of CO_2 or pick-up of oxygen

Air Evacuation

Older bottle fillers filled beer against CO_2 counter pressure, but, not surprisingly, it was found that the bottles' air content was too high due to the air in the bottle not being removed. Modern fillers have a vacuum system that evacuates about ninety percent of the air in the bottle before counter pressuring, which reduces potential air pick-up by a factor of nearly 10. Most recent fillers do this operation twice to get the air contents down to levels of about 20–40 ppb (parts per billion).

Filling

Filling the bottle as it rotates on the filler head takes up about half the available time and there are essentially two different ways of doing this—long tube and short tube.

The older method of filling is with a long tube that descends almost to the bottom of the bottle. Beer is run into the bottle from the filler bowl under the CO2 counter pressure of about 1–1.3 kg/sq. cm and residual gas goes out by way of a vent tube near the top of the bottle. This long tube method is relatively slow because of its length and the diameter being restricted (10–12 mm), but oxygen pick-up is quite low because of the quiet filling conditions with the submerged tube.

Fillers with short tubes give a greater throughput because the tube is for venting displaced gas, and beer goes down the outside of the tube. To avoid turbulence, the tube has a conical section on the outside and this deflects beer so that it runs quietly down the bottle walls.

Crowners

Crowns should be applied to bottles as quickly as possible after filling, to keep air contents down and prevent loss of beer. As a result, it is usual for crowners to be situated close downstream from the filler, and they are frequently integrated into the filler bloc to get full synchronization of the two operations.

Use of Antioxidants

Addition of ascorbic acid (vitamin C) as antioxidant is possible for preservation. Primarily, it is used to protect the color. Antioxidants can be used according to legal regulations.

A certain concentration of vitamin C gets lost even during the storage phase due to oxygen inside the packaging and dissolved in the product. This loss needs to be considered in the formulation. Higher additions are thus necessary, which need to be determined in advanced in stability and load tests.

Use of Preserving Agents

The corresponding statutory threshold values for permitted acids are in force for the preservation of foodstuff.

Effective conservation of sugar-sweetened, beer-based mixed drinks with preserving agents may not be sufficient and should be verified accurately. The conservation of raw materials is advisable since the microbiological stability can just be ensured during fractional removal in the bottling factory.

The most effective method for stabilizing the final beer is pasteurization of the filled bottles. It needs to be taken into account that with high sugar concentration, higher pasteurization unit (PU) values need to be applied than common in beer production. Due to the increased thermal load, the CO_2 concentration of the beer needs to be adjusted because of high pressure in the bottles. This can also have an influence on the filling level. It might be necessary to change the parameters usually used during beer filling.

Thermal Processes-Pasteurization

The heat treatment of food and drink, in order to kill spoilage organisms, dates back to the groundbreaking work of Louis Pasteur in the 1870s. He found that heating beer to temperatures between 50°C and 55°C was sufficient to preserve it. Within a few years, the term 'pasteurization' was coined and a number of food and drink industries adopted heat treatment as a means of preserving their products and this is still standard practice today. Not all materials respond to the same temperatures. Milk and some other foods need to be heated to very high temperatures. Beer is fortunate in not containing any pathogens and the nature of the common spoilage organisms is such that temperatures of about 60°C will suffice.

In beer, the common contaminants are residual pitching yeast, wild yeast, lactic acid bacteria, acetic acid bacteria, and cocci. Of these, lactic acid bacteria are more heat-resistant than most, with acetic acid bacteria the easiest to kill. Occasionally, heat-resistant wild yeasts emerge but this is not a problem in a clean plant with good yeast management. Other factors in the brewer's favor are the alcohol and CO_2 content of beer, both of which assist in killing off spoilage organisms. In addition, a well-filtered beer will have a very low microbial loading and this helps in reducing the heat requirements. Lack of oxygen and the presence of hop compounds also help.

Theory of Pasteurization

The basis of pasteurization is establishing the minimum time and temperature required to destroy all expected biological contaminants at the highest concentrations at which they may occur in filtered beer. Different food products have different requirements for pasteurization, and those that can contain spore-forming bacteria require much higher heat treatment than beer. Mixed populations of common brewery contaminating organisms were subject to a range of times and temperatures in beer (Figure 10.1—known as a lethal rate curve) and were examined for subsequent viability. Typically, at temperatures of over 50°C, an increase in temperature of 7°C accelerated the rates of cell-kill by ten times. Therefore:

- 53°C: minimum time to kill population is 56 minutes
- 60°C: minimum time to kill population is 5.6 minutes
- 67°C: minimum time to kill population is 0.56 minute

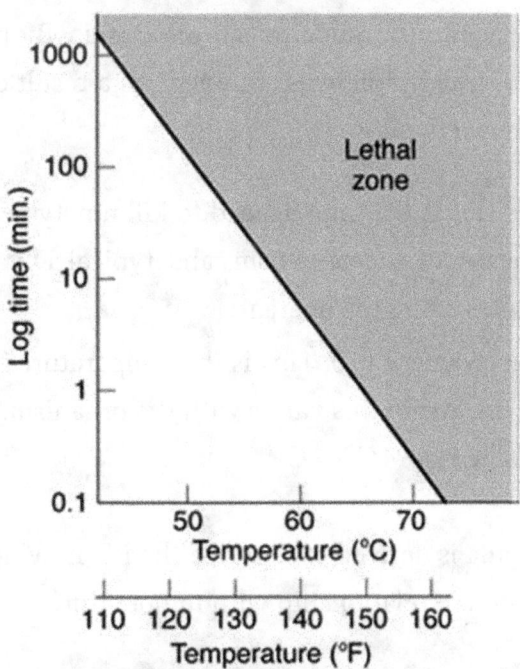

Figure 10.1: The effect of time and temperature on the viability of a mixed population of yeasts and brewery bacteria. The hatched area shows the range of conditions where all cells are killed (*Hough et al., 1982*)

Pasteurization Unit

One pasteurization unit (PU) for beer has been arbitrarily defined as the biological destruction obtained by holding a beer for one minute at 60°C. Therefore, in Figure 10.1, the point at which the line crosses the 60°C line gives the thermal resistance of the particular suspension of organisms. This is 5.6 minutes—and so, to achieve effective pasteurization, the holding time at 60°C must exceed 5.6 minute. The slope of the line in Figure 10.1 is known as 'Z value'. The lethal effect (PU) is simply the product of the lethal rate and the time of application. The lethal effect at various temperatures in a process is additive, therefore the sum of the lethal effect is the quantity of sterilization achieved:

Lethal Effect = L × t (PU)
Where,

    L = Lethal rate

    T = Time held at temperature T°C

    $L = 1/\text{Log}^{-1} (60 - T/Z)$

In practice, heat treatments aim to give at least 5 PU and 10 to 30 is normal. Until the 1950s, the time-temperature relationship was not well understood, and brewers pasteurized their beer largely based on past experience. Around this time, a number of researchers started to look at different spoilage organisms and the effect of various temperatures. It was found that there were differences between organisms, as mentioned previously, and that the temperature effect was exponential, which is not surprising. The effect was not quite the same for all organisms. By plotting a graph of logarithm of time against temperature, a straight line was obtained. As a result of this work two main factors were devised to describe the heat effect:

1. *Decimal reduction time:* This is the time needed to kill ninety percent of organisms at a given temperature. This varies with the organism, and typical D-values at 60°C are one to five minutes, with two minutes being the average.

2. Temperature dependence value Z (°C). This is the temperature increase needed to reduce the D-value by ninety percent. Again this varies with the organism. Values range from 3–8°C and the average accepted value is 6.94°C (~7°C).

From these two factors a formula is derived to describe the PU in which 1 PU is defined as holding beer at 60°C for one minute, and the relationship with temperature is:

$$\textbf{No. of PU} = \textbf{1.393}^{(T-60)} \times \textbf{time (min)}$$

The numbers of PUs delivered at a range of temperatures over a one-minute period are shown in Table 10.1.

From the figures in Table 10.1, it can be seen that a twenty-minute hold at 60°C will deliver twenty PUs to a packaged beer, while twenty seconds at 72.5°C will do the same for bulk beer going through a flash or plate pasteurizer.

Table 10.1: The Number of PUs Delivered at a Range of Temperatures over a 1-min Period

Temp. (°C)	PU
60	1
62	1.9
64	3.7
66	7.2
68	14
70	27
72	52
72.5	62

In practice, pasteurization falls into two categories—flash pasteurization and tunnel pasteurization. Flash pasteurization is always used for keg beers (unless sterile filtration is used), as it is not possible to pasteurize such large containers (although it has been tried!). Flash pasteurization is also used for some bottle and can filling operations and is needed when filling pressure-sensitive containers, such as PET. This technique has the advantage that the capital cost of installation is not particularly high and the plant does not take up a lot of space.

Tunnel pasteurization is used on bottles and cans, and it is the most reliable way of producing products with a long shelf life in these packages, as all parts are treated. It relies on a lower temperature than flash pasteurization spread out over a longer time (up to one hour) because of the time taken for heat to penetrate the package.

The effect of temperature on beer flavor is not entirely clear, apart from the fact that high oxygen levels and pasteurization do not go together. A stale, bread-like flavor is often the result. By keeping dissolved oxygen levels to a minimum, and by using only modest levels of pasteurization, it is possible to produce beer that can remain commercially acceptable for weeks or months. There is also some controversy over whether flash pasteurization is more or less damaging to flavor than tunnel pasteurization. Over-pasteurization by either method is deleterious.

Tunnel Pasteurization

Tunnel pasteurization is similar to flash pasteurization in that it involves heating the package to the correct temperature, holding at that temperature, and then cooling down. The timescale is very much longer, however—up to one hour—and the peak temperature achieved is lower (at about 60˚C). There are a number of reasons why this long timespan is needed. First, the rate at which heat is conducted through a container wall and then through the contents is quite long. There is a 'lag' of about ten minutes in this heating process. Second, with bottles, a rapid rise in temperature would cause thermal stresses that could result in the bottle bursting. Third, there is a steep rise in pressure when a highly carbonated package is heated and again there is a risk of bursting (Figure 10.2).

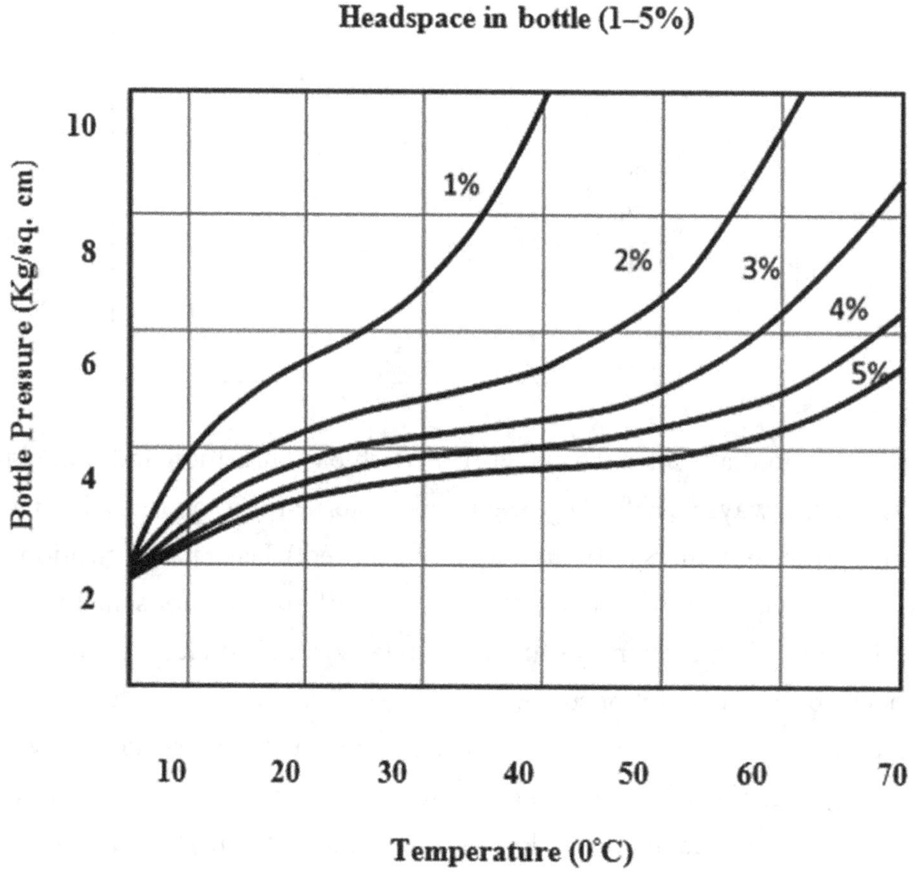

Headspace in bottle (1–5%)

Figure 10.2: Relationship between headspace and bottle pressure

Bottles have a wide range of failure pressures, and cans are specifically designed and manufactured to withstand only 6 kg/sq. cm. For these reasons, a low temperature–long time profile is the only practical option to achieve the desired PUs (usually about ten) evenly distributed throughout the container.

Tunnel pasteurizers are very large equipment. Each one consists of a very long enclosed chamber where cans or bottles are fed in at one end on a conveyor, heated and cooled as they travel through, and emerge from the other end. Frequently, it has two decks to save space. The two main components of the pasteurizer are the water spray and circulation systems and the package transport system.

The Water Heating and Spraying System

Cans and bottles are very slowly moved through the pasteurizer, and heating is achieved by spraying warm water onto them. These spray sets are in zones across the length of the machine and are at set temperatures. The first zone, for example, will be at a temperature of, say 20–22°C, to gently warm the container up to about 9–10°C. The water falling past the containers and through the conveyor is collected in a trough underneath for reuse. Water at progressively higher temperatures is sprayed onto the containers to bring them up to 60°C.

The most critical section in the pasteurizer is called the superheat zone, which is the last heating zone before the holding section at 60°C. The temperature in the superheat zone must be very accurately controlled at 61–65°C to ensure that the containers are brought up to the correct temperature. Each machine is slightly different, and this temperature needs to be established on initial commissioning to give the required PUs. It will also differ between bottles and cans. It is important, if there is a stoppage in the pasteurizer that the containers do not get overheated if trapped in superheat zone. This can lead to over-pasteurization and burst containers, especially cans, if the temperature goes too high.

The most modern pasteurizers have very sophisticated controls and can compute the number of PUs cumulatively as the containers travel through. This can be used to adjust the temperatures in the event of a stoppage. A more conventional way of checking a machine's performance is to feed through a recording thermometer. There is usually a filled dummy bottle or can on a base plate with an accurate temperature probe inside, which is connected up to a recorder. It is normally inserted into the pasteurizer, then retrieved at the back end and downloaded. This gives more realistic information as it tells you what is going on inside a container. A typical temperature–time profile for a seven-zone pasteurizer is shown in Table 10.2.

Table 10.2: Typical Temperature–Time Profile for a Seven-Zone Pasteurizer

Zone No	Spray Temp (°C)	Spray Time (min)	Package Temperature In	Out
1 Pre-heat	22	6	2	9
2 Pre-heat	32	7	9	21
3 Super-heat	65	14	21	60
4 Hold	60	6	60	60
5 Cool down	40	10	60	43
6 Cool down	32	7	43	36
7 Cool down	22	6	36	28
Total: 56				

In order to save on water and energy, there is usually a complex system of pipes running backward and forward between the different zones. One can see from the above example that the first and last zones work at the same temperature, as do zones 2 and 6. As the cooling zones will pick up heat from the warm containers, this water is pumped to the front end where it warms up the cold incoming packages. It will lose heat as a result, and is then pumped to the back end to cool down more containers again.

However, each zone should have facilities for heating up and cooling down its reservoir to cater for start-up and shutdown conditions as well as occasional stoppages. If the pasteurizer is in equilibrium then only the superheat zone needs significant steam input, the others need only a modest top-up. The energy consumption is still quite high with a tunnel pasteurizer, and fifty percent recovery is the best that can be achieved because the containers go in cold and come out at around 25–30°C. Most modern pasteurizers have more than the seven zones mentioned previously in order to improve temperature control and utilities use. If a pasteurizer does get out of balance, for example, when being emptied, then water consumption will rise due to the heat imbalance and overflow of the troughs as cold water is added. Conversely, steam consumption is high when filling.

The pressure generated in heated containers was also mentioned, and this is illustrated Figure 10.2. One can see from this that control of CO_2 in-package becomes important to keep pressures down and control of contents is another big factor. The only compressible part of a package is the gas headspace and if this is too small the pressure will rise more steeply than in a container with average contents.

The Water System

Pasteurizers are prone to corrosion and slime growth due to the warm damp conditions and it is necessary to add inhibitors to control both of these. Another way to suppress slime is to circulate hot water through all the sections. The use of inhibitors with cans needs to be done with care since the decoration occasionally suffers if the concentrations are too high, and the cans would be unfit for sale.

Blocked spray jets are a distinct quality hazard in pasteurizers as they may lead to under-pasteurization. This occurs due to scale or slime build-up and is best countered by regular inspection of the water tanks underneath and by internal inspection of the machine. Cold spots due to blocked jets or other abnormal flow conditions can be picked up by using a traveling recorder and placing it at different points across the pasteurizer.

Flash Pasteurization

Flash pasteurization involves using a plate heat exchanger to rapidly heat the beer up to a temperature of about 70°C, holding it at this temperature for some seconds, and then chilling it down again for packaging. In practice, the plates in a pasteurizer are sized to give a substantial degree of heat recovery (ninety to ninety-five percent). When the incoming beer comes out of the regeneration section, it helps if it is close to its final temperature so that this final heating step can be better controlled. The holding tubes on a pasteurizer usually consist of an elongated spiral of 100- or 150-mm pipe in sections with narrower connections. The most practical holding time is generally twenty seconds, so to achieve 20 PU in the beer a target temperature of 72.5°C would be used. This time span allows good control without the need for huge holding tubes. Brewers seldom use less than 10 PU for flash pasteurization and twenty to fifty is more common.

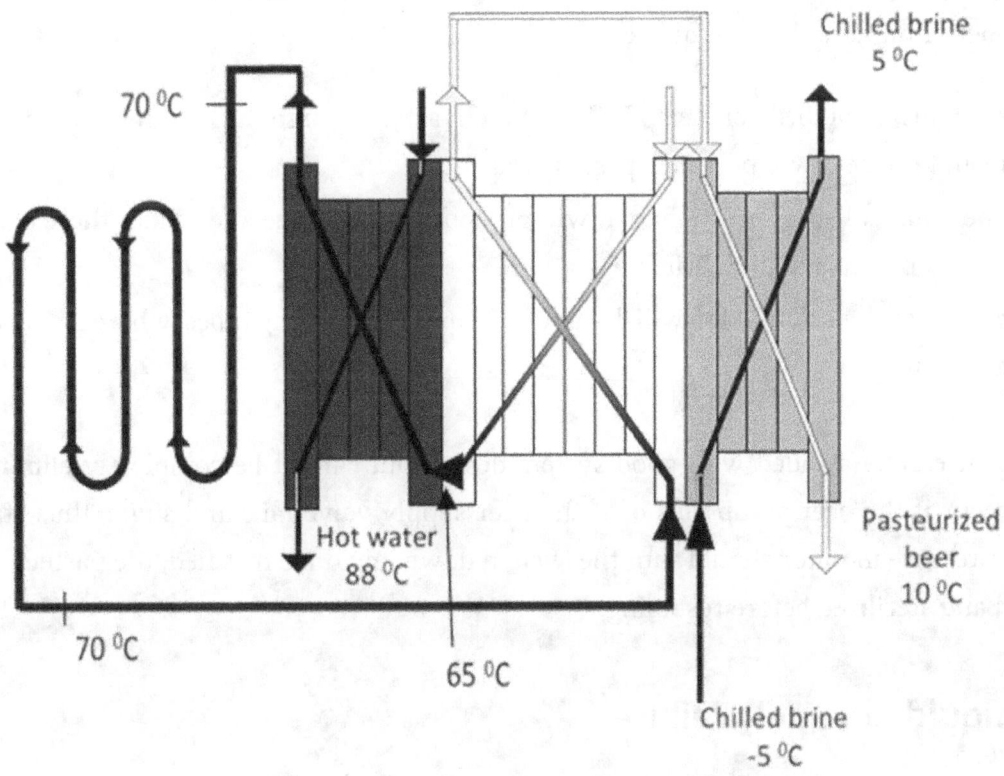

Figure 10.3: Schematic Representation of a Flash Pasteurizer

Quality Safeguards

There are a number of practical difficulties in operating a pasteurizer, which could become serious issues if allowed to continue. The all-important thing is to ensure that the beer is not under-pasteurized, and there are various failure modes where this could occur.

Temperature Drop

If the temperature coming out of the heating section of a pasteurizer is not high enough, incomplete pasteurization will result. This should be detected by a probe that would feed a signal back to the controller calling for more heat, and should immediately put the pasteurizer onto recycle mode, taking beer from the outlet and feeding it directly into the inlet until temperature conditions are restored. Usually, prolonged re-circulation is avoided by shutting down the pasteurizer if the fault is serious, such as a steam supply failure or pump failure on the heating loop.

Gas Breakout

When beer is heated to 65°C, the solubility of CO_2 drops so low that unless a high pressure is maintained in the pasteurizer there will be a breakout of gas, filling the holding tubes with fob. This has a number of unpleasant consequences.

1. The expansion of the beer means it flows much faster through the holding tubes in a turbulent state and the beer will be under-pasteurized

2. As the foam collapses on cooling, it will probably form a haze and, under these conditions, it is usually permanent and visible

3. There is also the risk that fob will dry and bake onto the holding tubes, where it may eventually become infected

Gas breakout can be avoided with good system design but cannot be completely eliminated. It is likely to occur if the beer pump fails or if the beer supply valve fails and shuts, thus starving the system. A pressure monitor should shut the system down and once rectified, the pasteurizer should be cleaned and sterilized before resuming.

Plate (Liquid Channel) Failure

A general cause of pasteurizer failure is leakage through one of the liquid channel plates. As shown in Figure 10.3, the pressure on the beer inlet side is always going to be higher than that of the pasteurized beer coming back on the other side of the liquid channel plates. Pasteurizer plates occasionally fail due to stress corrosion. This allows unpasteurized beer to short-circuit the pasteurizer, frequently leading to failed kegs in trade. Detecting this problem is not easy. The usual routine is to pressure-test the plate pack on a weekly basis.

The solution is to install an additional pump in the pasteurizer at the end of the holding tubes, the inlet to the regeneration section. This means that the hot pasteurized beer is boosted in pressure as it starts to cool down and exchange its heat with the incoming unpasteurized product. If there is a plate failure in these conditions then it will always be with pasteurized beer getting out, and not raw beer getting in.

Flash pasteurization equipment is mechanically very simple, cheap to buy and easy to operate (Figure 10.3). It is the ideal system for filling kegs and is becoming more widely used for small packs such as PET and glass bottles. The key to success in this later application is to ensure that the bottles, caps and filling equipment are sterile.

Table 10.3: Sterilization and Pasteurization Processes in the Brewery

Product	Process	Incubation	Comments
Beer	Flash pasteurization, (flow pasteurization with heat storage)	Vegetative cells, beer pests	Ascospores from wild yeast can survive
Beer	Hot filling (high-pressure filler)	Vegetative cells, beer pests	Cooling down necessary; ascospores from wild yeast can survive
Beer	Pasteurization ('pasteurization' in bottles or cans)	Vegetative cells, beer pests	Cooling down necessary; ascospores from wild yeast can survive

Packaging Line Efficiency

The measurement of a packing line's performance is an essential requirement in any plant, and this is usually accompanied by a system of analyzing and rectifying stoppages. This is frequently used to justify capital expenditure for replacing an item of equipment. There are two components to the overall performance of a line—machine efficiency and machine utilization.

Machine Efficiency (ME) is usually measured at the slowest component of the line and, in the case of bottling and canning operations, this is almost always the filler. If this machine is rated at, say, 250 containers per minute and it always runs at this speed, then its efficiency should be hundred percent. However, if there are, for example, gaps in the supply of empties or some process reason for slow running, the efficiency will drop.

Machine Utilization (MU) is the time that the filler is running in relation to the total planned time. If the operation is on two shifts per day of eight hours each shift, the available hours are ninety-six hours per week. It may not be busy enough to use the whole ninety-six hours, but, say, only ninety hours. The rest is used toward week cleaning, briefing, training, and so on. If, during the ninety hrs of planned production, there are nine hours of downtime, machine utilization (MU) is:

$$\text{Machine Utilization (MU)} = \frac{90 - 13}{100} = 77\%$$

These two factors of efficiency and utilization are usually combined to give *Machine Effective Utilization* (MEU):

$$\text{MEU} = [\text{ME} - \text{MU}]$$

A typical figure for MEU would range between sixty-five and eighty-five percent. This is often governed not by machine stoppages but factors such as size and quality changes or type of pack.

Relative Machine Speeds

For bottling and canning lines the key item of plant is the filler and other machines should be rated relative to it. The best arrangement is where the items immediately before and after the filler are capable of running about ten percent faster and the next machines faster still. This is sometimes referred to as the V-Diagram (Figure 10.3) from the way in which the speeds look when plotted on a graph with the order of the machinery. A typical example is given next for a bottling line.

Equipment	Relative Speed (%)
Bottle Washer	110–115
Filler	100
Pasteurizer	110
Labeler	110–120
Case Packer	120–125

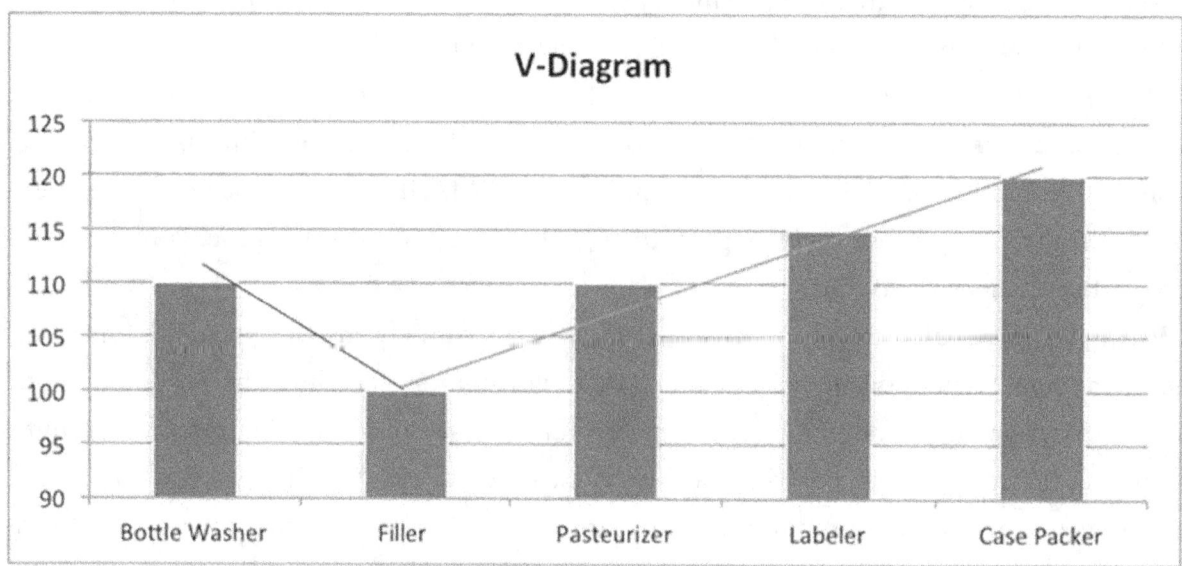

Figure 10.4 : Relative Machine Speed

This arrangement of relative speeds means that the line is in a state of compression before the filler with the conveyors usually full, but after the filler bottles are being taken away faster than they can be supplied. This makes the filler the pinch point on the line, but there are good reasons for doing this.

Firstly, the cost—the filler is probably going to be the most expensive single component of the line and it pays to maximize its throughput. The cost of uprating the filler by twenty-five percent would be much higher than for a case packer.

Secondly, the quality of beer—the filler performs at its best when there are no interruptions and should always have a supply of bottles to fill, and no bottlenecks downstream. If the filler is constantly stopping and starting, there is going to be more oxygen pick-up, variable carbonation, and more variable contents. The filler/crowner machine and the pasteurizer are the main machines on the line, where beer quality is directly affected, and they need to be protected.

In summary, it can be seen that beer packaging is of great importance and the most expensive part of the whole process. It has changed greatly over the last fifty years, and this will continue. Quality, economics, new technology, and the drive for increased market share and profit will keep this area in a constant state of change.

Warehousing and Distribution

Finally, it is important that the beer be distributed to the consumer in top condition. Today, this is an increasingly complex operation, and the supply chain is a key element of successful distribution. Nevertheless, we must remember that an excellent product in the brewery does not necessarily mean that it will still be excellent when it reaches the consumer and that large companies in particular have struggled because the beer is in the wrong place at the wrong time. So, paying careful attention to logistics and supply chain is essential to both the large international and small breweries.

Conclusion

Packaging is a vital part of the brewery operation. The rise in the number of people who are drinking beer at home and the influence of retail supermarkets has meant that effective packaging of beer in bottles and cans is essential to catch the eye of the purchaser. Huge amounts of money are being spent by multinational companies (MNCs) on packaging developments, and certainly much more than is spent on research into processing or raw materials.

In many parts of the world, the major package remains the returnable bottles. However, local breweries are in competition with the breweries of international brands putting their traditional markets under threat by more sophisticated packaging. This has resulted in a raising of standards of packaging to preserve the market for local beers.

Brewing is increasingly becoming a trans-national business and it is likely that as markets develop, the trend will continue towards packaging in non-returnable bottles, probably made of plastic. This will increase the pressure to find effective ways of recycling the empty package.

11

Microbiology and Microbiological Control in the Brewery

Introduction

Beer is microbiologically stable and, therefore, not subject to the numerous spoilage microorganisms that colonize most foods or non-alcoholic beverages. It has been subject to exhaustive yeast growth and, therefore, like other fermented foods, is largely resistant to further microbial development. The reasons for this are several:

1. The low pH (around four) inhibits most microorganisms
2. The high alcohol concentration is toxic to many microorganisms
3. The antiseptic action of hop α-acids is bacteriostatic to many bacteria, particularly gram positive types
4. Only residual nutrients (pentose sugars, higher maltooligosaccharides) are available as carbon sources

Despite the factors limiting microbial spoilage, there are various yeasts and bacteria that can grow in beer, particularly if the storage conditions are poor and oxygen is allowed access. Fortunately, none of these organisms is pathogenic, so the only problem for the brewing microbiologist is consistency of the appearance and organoleptic qualities of the final product.

The nature of brewing is rapidly changing—smaller breweries are giving way to the larger, more modern facilities. Some of the older, larger breweries are forced to undergo change to meet modern practices and demands. Cleaning and disinfecting methods are being modified; detergent manufacturers are constantly improving products designed for current operations.

When the brewer is faced with problems in cleaning and disinfection of the tanks, lines, equipment and surroundings, new methods or new materials for dealing with cleaning and disinfection must be developed in tandem with the operation. Although wild yeasts, wort, beer spoilage bacteria, and molds are still present in the raw materials, air, water and equipment, the use of closed-circuit cleaning disinfection with custom-made detergents and sanitizers gives the brewer a better chance in combating these biological agents.

Another area of beer production that creates great concern for the microbiologists is aseptic filling of packages after microbiological stabilization by heat (for example, pasteurization) or by microfiltration. The operation must include:

1. Sterile bottles or cans
2. Sterile beer
3. Sterile transfer lines
4. Sterile filler
5. Sterile package closing facilities

This operation requires the utmost cooperation between the microbiologist, production engineer, and operator, coupled with easily cleanable equipment and careful cleaning and sanitizer application.

Microbioorganisms Encountered in Brewing

A. Yeast

a. Pitching or Production Yeast

Pitching yeast is the only microorganism deliberately included in the fermentation or lagering process. The yeast cells become undesirable only after lagering, when they must be removed from lager or bright beers. Since pitching yeast represents a biological population, a certain population of cells will be dead or dying during fermentation. The microbiologist can determine the percentage of dead cells in a population, thus providing an index of yeast health for the brewer. Also, since yeast cells propagate by formation of buds, a bud count can be another feature of yeast vigor (in rapid fermentation, it is not unusual to observe that ninety-five percent of the yeast cells are budding). The dead portion of a yeast population is subject to autolysis in which the cells break up and the cellular contents are extruded to the fermenting beer. Constituents of the cell contents include proteins, fats, and carbohydrates, which can serve as nutrients for the beer spoilage bacteria.

b. Wild Yeast

Any yeast other than pitching yeast is considered as wild yeast. If the wild yeast gains entrance to the wort or beer and exhibits a growth rate superior to production yeast, there is a danger that, after many generations, the pitching yeast population would be in minority and eventually disappear. In case the wild yeast exhibits an equivalent growth rate to pitching yeast, then off-flavors may develop in the beer, or the wild yeast may interfere with the normal flocculation pattern of the pitching yeast, resulting in hung fermentation. The practice of acid-washing yeast to remove bacteria is not effective against wild yeasts.

a. *Aerobic Yeast*: These yeasts require air to grow and metabolize. Members of this group can be found in soil, water, and air sample and in areas where wort, beer, or sugar-bearing liquid occurs. Cracks in floors or valve, and so on, may create focus for wild yeast growth. Fortunately, these yeasts do not pose a threat to brewing because they do not grow under fermentation or lagering conditions.

b. *Anaerobic Yeast:* These yeasts, like production yeast can grow and metabolize in absence of air.

The genus *Saccharomyces* includes many types of yeast that have different metabolic and generic pattern. For example, *S. cerevisiae* is known as baker's yeast and is capable of growing in the absence or presence of air. *S. uvarum*, the lager yeast, has lost the ability to grow in the absence of air and gets its energy by fermenting sugars in the absence of air. Yeasts similar in growth and metabolic activity to pitching yeasts can be a problem to brewing.

Saccharomyces pastorianus is one such yeast and can grow in wort and produce off-flavor in the product. *Saccharomyces diastaticus* is another potential contaminant and has been included in cases of super attenuation in beer because this yeast can metabolize starch fragments. Wild yeast usually cannot be detected microscopically because of similarity to pitching yeast and the fact that they usually occur in small numbers in proportion to pitching yeast.

A recent observation in fermenting tanks has been the killer yeast phenomenon. Apparently, certain types of yeast that are capable of growing in wort under fermentation conditions have been designated killer types because they produce a toxin lethal to pitching yeast. Not all strains of pitching yeast are susceptible to this infection.

B. Molds

There are many species of mold, and they are capable of growing on damp surfaces at cellar temperatures. Mold growth usually consists of chain-like structures called 'mycelia', which link to form visible masses such as described as mold pad in the bottoms of empty bottles returned from the market/trade. These molds may also grow on flat surfaces, such as walls or ceilings, to create visual nuisances. While molds do not pose a direct threat to beer spoilage, the unsightly presence of these organisms creates cleaning problems for the brewer. Molds are known to grow on the thin water layer that exists on the surface of the syrups or high sugar concentrated solutions. Brewers who use syrup adjuncts find ultraviolet light sources situated on tank tops to be effective in preventing mold growth.

Molds such as *Aspergillus flavus* are capable of producing a class of poison called 'mycotoxins', which are known to cause poisoning in animal feed-infected grains. As a general rule, brewers should reject moldy barley and avoid problems caused by infected grain.

Mold growth on the walls and ceiling can be combated with conventional cleaners-antifungal paints are applied where need for constantly damp surfaces. Some molds produce spores that are microscopic and, when released into the atmosphere, tend to cause mild allergic reactions to susceptible personnel.

The observation of the mold *Geotrichum candidum* in food processing plants is generally designated 'machinery mold' by sanitation personnel and treated as an index of poor housekeeping, particularly in the packaging area. The mold is not harmful to man but its presence is not desirable in packaged foods and so on.

Molds are not heat-resistant microorganisms and do not survive during kettle boil; they are not usually found in beer.

C. Bacteria

When the wort leaves the boiling kettle, very few bacteria remain alive. Some bacterial spores survive boiling but these do not develop in fermentation to cause spoilage problems. Beer is a poor medium for wort bacteria for the following reasons:

- Low fermentation, lagering and finishing temperatures in lager beer production do not favor growth of contaminating bacteria
- Low oxygen tension suppresses growth of most aerobic bacteria types after the first few hours in fermentation when yeast rapidly depletes oxygen entrained in wort
- Nitrogen sources such as amino acids, peptides needed for the bacterial growth are also rapidly scavenged by the mass of pitching yeast in the fermenter
- Low pH—as fermentation continues, yeast metabolic activity lowers the pH of the beer below the range that supports optimum bacterial growth
- The formation of ethanol, fusel alcohols, and esters, as well as the presence of hop constituents, may combine to act as growth suppressants to bacteria

Thus, while the brewing environment is not conducive to optimum bacterial growth of most species, some do survive to cause problems.

A. Wort Spoilage Bacteria

Instances of wort spoilage are due to infestation of the wort after cooling. In order to effect spoilage by contaminating organisms, the wort would need to stand for a considerable period of time in absence of pitching yeast before the contaminants could act.

Some of the known wort spoilage bacteria are briefly described and include:

- *Obesumbacterium proteus (Hafnea protea)*: Often found as a contaminant of pitching yeast, it exists as a gram negative. *Obesumbacterium proteus* grows in wort at pH 4.3 and typically produces dimethyl sulfide, which has a distinct creamed-corn aroma in beer.
- *Escherichia species* will grow in un-pitched wort to produce vegetable-like flavor, but will not grow in beer. Their presence is an indicator of an incursion from outside sources.

- *Bacillus species,* which are gram positive rods capable of forming heat-resistant spores that may survive the kettle boil but do not persist in fermenting wort. Grain materials are usually the source of these bacteria.
- *Pseudomonas species* are gram negative rods. Occasionally found associated with grain or untreated water, they may be airborne through dust but do not grow in beer.

B. Beer Spoilage Bacteria

Bacteria capable of causing beer spoilage during production must have ability to grow and metabolize under anaerobic conditions.

There are two types of within the family of *Lactobacteriaceae:*

a. *Lactobacillus* species are gram positive, non-spore-forming rods that are capable of growing in beer to produce visible haze as well as flavor-depressing by-products.

b. *Pediococcus* species are gram positive coccus (round) cells occurring as one, two, or group of four cells (tetrads). These bacteria are the most important from the brewing standpoint, because they can grow and produce very distinctive off-flavor in beer.

Beer may be returned from the market/trade because of these bacteria:

a. *Zymomonas Species:* These bacteria are gram negative rods capable of at least limited growth in beer. They produce hydrogen sulfide and, in some case, acetaldehyde.

b. *Acetobacter Species:* These are gram negative bacteria that grow aerobically and typically oxidize ethanol to acetic acid as a source of energy. They are usually not encountered in production but often are found in spoiled beer from opened packages.

C. Algae

Algae are plants ranging from a single cell to many yards in length (such as seaweeds). They are chlorophyll-bearing and, as such, require sunlight for growth. These plants are unknown to the brewing process but their presence may be observed in bottle rinsers or pasteurizers, where algal filaments contribute to slime formation and blockage of the spray nozzle. Elimination of these organisms is usually achieved through a routine cleaning procedures.

D. Viruses

Viruses capable of infecting brewer's yeast have been found. Incursion of living viruses from water supplies and so on is a possibility in brewing but detection would be beyond the scope of the brewing microbiological service. At present, there appears to be no problem of viral infection in brewing.

12

Cleaning and Disinfecting

THE PURPOSE OF cleaning is to permanently remove all the soil from the surfaces of the plant and to leave it in a condition suitable for use.

The purpose of sterilizing is to kill any microorganisms that remain on the internal surfaces of the plant after cleaning, so that the wort or beer is not subsequently contaminated.

Detergents help the cleaning process by:

- Penetrating the soil, usually by increasing the wetting power of the cleaning liquid
- Dissolving the soil
- Dispersing the soil and holding it in suspension so that it does not re-deposit
- Carrying the soil away as the cleaning liquid is rinsed off

Sterilant work is done by: Creating the conditions of temperature, pH, chemical or surface activity that can destroy (kill) microorganisms.

Detergents

Wetting Power:

Water is always used as the medium for carrying the detergents used in cleaning brewing plants. Water has a relatively high surface tension—that is, it forms 'beads' on a surface rather than wetting it.

Most detergents contain a substance that reduces surface tension, and thereby increases the detergent's wetting power.

'Bead' of water sitting on a surface

Water with a 'wetting' agent added

Wetting agents have the tendency to foam so they may be supplemented with some form of anti-foam.

Dissolving:

When a substance is dissolved, it is chemically bound to the liquid, and the liquid is usually clear. If soil can be dissolved in the detergent liquid, it can be removed from the plant surface, and carried away easily as well.

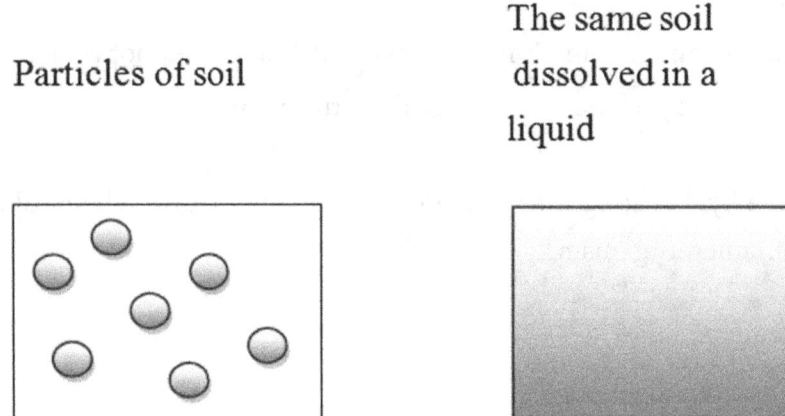

Particles of soil

The same soil dissolved in a liquid

There are two main types of soil that need to be removed from the surface of a brewing and packaging plant:

- Organic soil, which includes yeast, protein, fat and sugar; a plant that has a lot of organic soil should be cleaned with a detergent that contains compounds that can dissolve it. Alkalis like caustic soda dissolve organic soil, and caustic solutions are often used to clean fermenting vessels and brewhouse plants.
- Inorganic soil that includes scale or 'beer stone': Plants in some breweries become scaled up quite quickly, especially in areas with hard water. These plants need to be cleaned regularly with a detergent that dissolves scale. Acids like nitric acid or phosphoric acid are good at dissolving inorganic soil. Sequestering agents that can be added to alkaline detergents are also capable of dissolving scale.

Dispersion:

Not all the soil in brewing and packaging plants is soluble, though insoluble soil can be removed if it is 'dispersed' so that it can be carried away by the liquid.

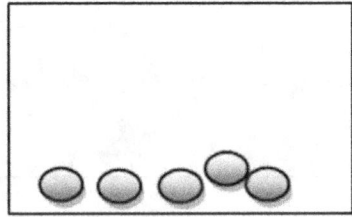

Soil on the Plant's surface

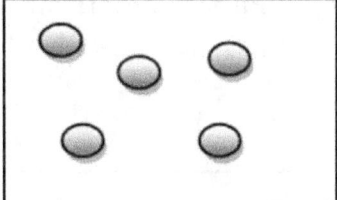

The same soil dispersed in a liquid

Detergents contain substances that help to disperse the soil, and to hold it in suspension so that it can be rinsed away.

Rinsing:

It is important that, at the completion of a cleaning cycle, no detergent and accompanying soil remains on the plant's surface. In other words, the detergent must be 'rinseable'.

Thus, to be effective, a detergent must be capable of adhering to the plant surface being cleaned—when the job is done, it must be rinsed away. Rinsing agents are added to the detergent to enable these two incompatible actions to take place.

Detergents are 'built' up from a number of constituents. The table below gives details of the most common constituents and their contribution to the effectiveness of the detergent:

Constituent	Effects	Benefit/Problem
Caustic soda.	Dissolves organic matter. Sterilizes especially when hot.	Does not rinse well. Very hazardous and cannot be used by hand. Dissolves aluminum. Denatured by CO2. Spraying a tank containing CO2 with caustic can create a vacuum and collapse the tank.
Other alkalis. e.g. silicates.	Dissolves organic matter.	Less aggressive than caustic soda. Very good dispersants.
Oxidants. e.g. Hypochlorite.	Help dissolve protein. Sterilizes.	Very corrosive unless at high pH.

Constituent	Effects	Benefit/Problem
Phosphates.	Soil removal.	Very good rinsing properties.
Acids. e.g. nitric, phosphoric.	Dissolves scale.	Corrosive in high concentrations. Not denatured by CO2.
Wetting agents. e.g. teepol.	Reduces surface tension.	May cause the detergent to foam.
Sequestering agents. For example, EDTA.	Prevent the formation of scale.	Expensive.

Temperature:

The temperature that a detergent operates at influences its effectiveness.

The action of caustic soda, for example is much more powerful at higher temperatures. The graph illustrates how caustic soda reaches maximum effectiveness at 85˚C.

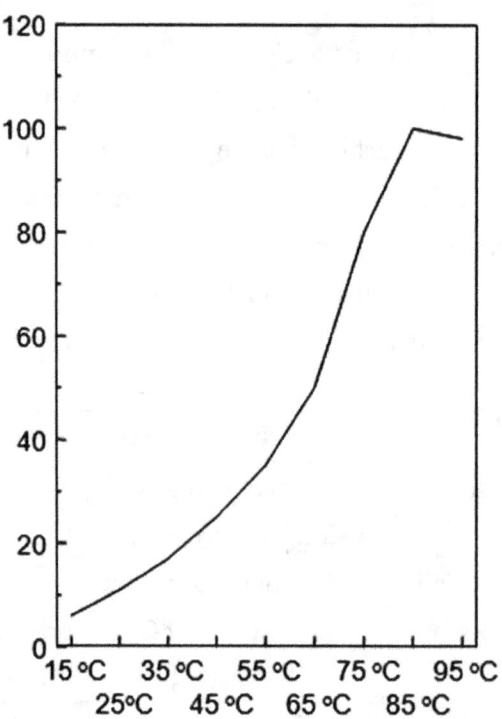

In the brewery, cleaning and disinfecting are of utmost importance when it comes to the quality and shelf life of the product. Comprehensive and regular cleaning and disinfecting measures, which must always go hand-in-hand, are therefore an important requirement for proper process management.

Necessary measures should be carried out, based on a fixed schedule that takes the type of product and operational anomalies into account. Cleaning and disinfecting have two different goals:

1. Product—contaminating substances are to be fully removed
2. Microorganisms are to be rendered inactive

Basically, contamination can be divided into two types:

1. <u>Inorganic contamination, which includes</u>:
   - Rust
   - Metal grit
   - Glass grit
   - Dust
   - Water scale
   - Beer scale
   - Soil (inorganic part)
2. <u>Organic contamination, which includes</u>:
   - Beer residues
   - Hop resins
   - Residue (organic parts) of cleaning and disinfecting agents
   - Fibers
   - Body greases
   - Soil (organic part)
   - Bacteria
   - Yeast
   - Molds

Cleaning

Cleaning is the removal of contamination or undesired residues from hard surfaces with the aid of chemical and/or physical cleaning methods and agents. Factors for successful cleaning include:

- Temperature (hot cleaning, cold cleaning)
- Cleaning time (the longer the cleaning time, the greater the cleaning success)
- Mechanics (pressure, volume flow, flow speed)
- Chemical (type and concentration of the cleaning agent)

The most important factor is the selection of a suitable cleaning agent. The following minimum requirements must be met by a suitable cleaning agent:

- Quick and complete solubility in water
- Quick swelling and detachment of the specific main components of the contaminant (such as protein, beer scale, hop resins and fur)
- High soiling—carrying capacity
- Easy rinsing, so as to reduce the final rinsing times
- Non-foaming or foam—reduction property for other foams
- Compatibility with materials used in production systems

An important requirement for effective cleaning is the ability to determine the concentration of the cleaning agent solution. This occurs either via simple titration or, in the case of automatic CIP systems, the concentration is determined via conductivity.

Mechanics are the physical conditions required for cleaning. These include the pressure (minimum 3–5 kg/sq. cm), volume flow and flow speed (3–4 m/s). These technical requirements must be taken into account during the planning and installation of systems. The temperature of the cleaning agent depends on the technical options, stubbornness of the contamination to be removed and, of course, chemical composition of the cleaning agent solution itself.

The following temperature ranges are common:

- Fermentation tanks, storage tanks and bright beer tanks: cold to 40°C
- Brewhouse, Lauter Tuns, mash tuns, wort coolers, all pipelines: 70–90°C

The chemical processes of breaking up soiling (for example, dissolving stone via acids), swelling of soil (for example, with dried-on starch and protein residues), saponification (for example, with grease), dispersion (for example, hop resins) and rinsing clean are subject to rate laws. Strictly speaking, the soaking time means the contact time of the cleaning agent with the soiling at the right concentration and temperature.

Within certain limits, the temperature, time, mechanics and chemical factors can be exchanged with one another. It is thus possible, for example, to compensate for a low cleaning temperature by increasing the cleaning agent concentration and/or the flow speed.

Cleaning Agents

Basically, cleaning agents can be divided into three large groups—neutral, alkaline and acidic. Cleaners within each group have different compositions (Table 12.1). Neutral cleaning agents are used when easily water-soluble, dispersible or emulsifying contamination is to be removed from surfaces. Alkaline cleaning agents are used when organic contamination that is not easily water-soluble is to be removed. In this case, a chemical transformation occurs—water-insoluble residues are changed to water-soluble fragments. Poorly water-soluble mineral contamination is removed when acidic cleaning agents chemically transform it into water-soluble salts.

Alkaline Cleaning Agents

Sodium hydroxide exhibits an excellent emulsifying capacity for protein. Accordingly, it is used in a wide variety of applications in breweries. Caustic potash has an even greater capacity for breaking up soiling than sodium hydroxide. It is only used in limited applications, however, due to the fact that is several times more expensive.

The disadvantages of sodium hydroxide include:

- No dispersion properties
- No surface-active effect
- No sequestering power
- Not easily rinsed out
- Attacks aluminum
- Foams at high pressure

To reduce these disadvantages to a minimum, various additives are added to the caustic soda solutions, depending on the requirements of the situation. Since alkaline solutions strongly tend to foam up, defoamers (anti-foaming agent) must be added to them. A point of difference between hot and cold defoamers is:

- Hot defoamers are used for bottle cleaning
- Cold defoamers are used as foam inhibitors in brewhouse and pipe cleaning

In many cases, more than just pure sodium hydroxide and an additive are used. In these cases, so-called prepared cleaners are used (that is, the user receives a prepared concentrate, which he must then dilute to a specific usage concentration).

However, Alkaline products are not suitable for use with aluminum. Special cleaning agents containing silicates must be used for aluminum system parts. Silicates are also contained in alkaline cleaning agents. They have very good cleaning, dispersion and emulsifying properties. Sodium meta-silicate is a very good corrosion inhibitor for aluminum. When using high temperatures, the silicate content in the cleaning agent may not be too high—otherwise, there is a risk that calcium silicate may precipitate. This is very difficult to remove.

Table 12.1 Cleaning Agents and Additive Substances

Component	Active Ingredient	Remarks	Use
Alkalis	NaOH/KOH	Removal of organic soil; carbonate formation via CO_2	CIP cleaning; bottle cleaning; keg cleaning; foam cleaning; floor cleaning
Acids	Phosphoric acid	Inorganic acid; removal of inorganic and organic soiling; release of phosphate in wastewater	CIP cleaning; keg cleaning; foam cleaning
	Nitric acid	Inorganic acid; removal of inorganic contamination (beer scale); passivation of stainless steel; release of nitrates in wastewater	CIP cleaning; keg cleaning
	Sulfuric acid	Inorganic acid; removal of inorganic contamination (beer scale); concentrate attacks stainless steel	CIP cleaning; bottle cleaning
	Hydrochloric acid	Inorganic acid; removal of inorganic contamination (hardness-mineral deposits); high corrosion potential	Stone removal systems; neutralization of wastewater
	Citric acid	Organic acid; removal of inorganic soiling	CIP cleaning; foam cleaning
Additive Substances			
Oxidizing cleaning booster	Hydrogen peroxide; chlorine	For the detachment of stubborn organic incrustations; products decompose during cleaning process	CIP cleaning; special cleaning in circulation process
Dispergators, sequestering agents	Polycarboxylates; phosphonates	Increase the soil-release capacity of the cleaning solution; improve cleaning success	Bottle cleaning; CIP cleaning
Surfactants; defoamers	Non-ionic surfactants	Decrease the surface tension of the cleaning solution; prevent disturbing foam-forming	Bottle cleaning; CIP cleaning
Complexing agents	EDTA; gluconate; phosphate	Complexing of the water-hardness and metal ions in the solution; prevent precipitation and deposits	Bottle cleaning; CIP cleaning; alkaline foam cleaning
Solubilizers	Alcohols; glycols	Stabilize individual ingredients in the concentrate of prepared cleaning agents	Prepared cleaning agents

Another ingredient of alkaline cleaning agents is sodium carbonate. It does not exhibit great cleaning, dispersion and emulsifying properties, but can be used as a component of the cleaning agent formulation for aluminum.

The primary purpose of the inclusion of complexing agents in alkaline cleaners is to prevent the formation of deposits on surfaces. They complex the builders of water-hardness so that they cannot be deposited on heat-exchanger surfaces. Previously, sodium polyphosphate was the most important of these, especially due to the fact that it also exhibits active cleaning properties. These types of complexing agents do not only sequester the minerals causing hardness dissolved by water, but can also dissolve mineral precipitations when used for a longer period of time.

Dispersion agents enhance the soil-release ability of the cleaning solution and prevent the growth of hardness—mineral crystals in water. They are advantageous in that they do not react with metal ions 1:1 as do complexing agents. Rather, they can be used in the sub-stoichiometric range (threshold effect).

Surfactants are usually water-soluble, surface-active chemicals added to cleaning agents as wetting and dispersion agents. Differentiation is made between anionic, cationic, amphoteric and non-ionic surfactants. Anionic and non-ionic surfactants primarily function as cleaners.

The advantages of using surfactants include:

- Surfactants allow the cleaning solution to penetrate into narrow gaps by lowering its surface tension
- Surfactants facilitate the penetration of the cleaning solution and thus accelerate the swelling of residue
- Surfactants emulsify grease in the aqueous phase
- Surfactants also transform water-insoluble soiling into an apparently soluble form

Acidic Cleaning Agents

The most suitable acidic cleaners are products with phosphoric acid. Phosphoric acid far exceeds nitric acid and sulfuric acid in its cleaning power. Almost all cleaning agents found on the market (except for pure beer scale removal agent) are based on phosphoric acid. Mixtures with other acids are also available, which both occurs for reasons of greater specific conductivity and allows for better dissolving of inorganic coatings. In conjunction with surfactants, products based on phosphoric acid are not harmful to stainless steel, even at greater temperatures (over 80˚C).

Mineral deposits (such as beer scale and water scale) are best removed with nitric acid cleaning agents. They transform water-insoluble salts into water-soluble, easily rinsing form. Nitric acid is not harmful to stainless steel. These products can also be used on aluminum. They are harmful to non-ferrous metals, however. The use of nitric acid products is not without risks, since the reaction with organic substances could release poisonous nitrous gases.

Disinfecting

Beer residue is the ideal nutrient media for every type of microorganism; they can also cause spoilage in beer. To prevent biologically induced spoilage, beer must be protected from contamination by germs. Sufficient protection is only achieved through hygienic production and packaging. Disinfecting agents used in the food/beer industry are tasked with ensuring that production equipment is free of microorganisms after use and subsequent cleaning. Microorganisms can be killed physically and chemically. Physical elimination involves heat treatment, UV and X-rays, and other methods. Disinfection through chemicals is possible via a host of disinfecting agents.

Efficient disinfection with any agent can only be ensured when a clean, physically intact smooth surface is being treated. All pores, cracks, deposits and other surface damages hinder meaningful disinfecting.

The requirements of disinfecting agents are as follows:

- Range of microbicidal effectiveness
- Effectiveness at low temperatures
- Toxicity
- Effectiveness under organic load
- Easily rinsed out
- Material compatibility (danger of corrosion)
- Stability when stored
- Environmental friendliness
- Economy

The following active disinfecting substances are important in the beverages industry (Table 12.2):

- Substances containing active chlorine
- Oxidizing agents
- Aldehydes
- Biguanides
- Quaternary ammonium compounds (QACs)
- Chlorine dioxide
- Halogenated carboxylic acids

Table 12.2 Disinfecting substances

Active Ingredient	Remarks	Use
Peroxyacetic acid	Acidic disinfecting agents with oxidizing effect (destroy cell membrane); conditional stackable due to loss of effectiveness; automatic dosing via inorganic conducive acids only; sealing materials may be harmful with extended contact; very broad range of effectiveness	Bottle cleaning; CIP cleaning
Hydrogen peroxide	Neutral disinfecting agent with oxidizing effect; very environmentally and wastewater friendly, since it decomposes with organic material in water and oxygen; high usage concentration; very broad range of effectiveness	CIP cleaning; spray disinfecting
Active chlorine (sodium hypochlorite)	Alkaline disinfecting agent with oxidizing effect; danger of chlorophenols formation (negatively effects taste of the product); very broad range of effectiveness; ATTENTION: when mixing with acidic solutions, chlorine gas is released	Bottle cleaning; CIP cleaning; drinking-water disinfecting
Chlorine dioxide	Disinfecting agent with oxidizing effect; two-component system that is mixed on-site when used; economical operating costs, but high investment costs; very broad range of effectiveness	Bottle cleaning; CIP cleaning; drinking-water disinfecting
Quaternary ammonium compounds	Neutral disinfecting agent (surfactants); destroys the cell membrane; heavily foaming (not suitable for CIP); surface-active; relatively difficult to rinse out due to the surface activity (adheres well to the surface)	Static disinfecting; spray disinfecting

Active Ingredient	Remarks	Use
Biguanides	Neutral disinfecting agent; destroys the cell membrane; forms deposits in alkaline medium; well suited to manual applications	Head space disinfecting; static disinfecting; spray disinfecting
Aldehydes	Neutral disinfecting agent; broad range of effectiveness and thus highly effective; oxidizes with air and then forms brown deposits	Static disinfecting; head space disinfecting
Halogenated carboxylic acids	Acidic disinfecting agent; very broad range of effectiveness; cannot be processed manually due to high toxicity; release AOX in wastewater; offered as a combined cleaning and disinfecting agent together with inorganic acid	CIP cleaning

AOX: Sum of all absorbable organic halogen substances

In the past, disinfecting products containing phenol derivatives were often used. Today, however, they are of very little importance in this field.

Products containing active chlorine have long been used to disinfect systems in the food industry. Generally, products based on sodium hypochlorite are used. In the weak acidic range, chlorine disinfecting occurs much quicker than in the alkaline range. For one thing, the decreasing stability of the chlorine carrier on the acidic level greatly reduces its usefulness. There is also a risk of pitting of chrome-nickel-steel and aluminum surfaces on the acidic level. It is absolutely imperative that the pH range be between ten and twelve.

If organic contamination is present, the effectiveness of the chlorine is reduced considerably (chlorine degradation). Chlorine products are also unstable at high temperatures (over 40°C).

Products containing peroxyacetic acid are increasingly being used for disinfecting in the food industry. Such products are advantageous in that they are practically undetectable in the case of insufficient final rinsing. Peroxyacetic acid is a strong oxidizing agent comprised of a mixture of peroxyacetic acid, acetic acid and hydrogen peroxide. Practically all microorganisms are killed with these products. Room temperature and a low concentration kill not only vegetative bacteria of all types, yeasts and molds, but also endospores of the otherwise difficult-to-combat *Bacillus* spp. and *Clostridium* spp. To a certain extent, a disinfecting agent solution containing peroxyacetic acid can even be stacked. It is imperative, however, that it be absolutely free of organic contamination here.

Quaternary ammonium compounds or QACs are cationic, surface-active substances that can actively penetrate gaps and pores thanks to their low surface tension. They are characterized by their effectiveness across a very broad pH range, and are effective in both the acidic and alkaline ranges. They are indifferent to metals and plastics and do not cause corrosion. At the concentrations used, QACs are not dangerous to handle, have no odor and are not harmful to skin. Gram positive bacteria are easily killed, whereas greater concentrations or longer contact times are required for gram negative bacteria in general. Quaternary ammonium compounds foam very heavily and can usually not be used in CIP systems.

Due to their substantivity, these products adhere strongly to the surface, which makes the products very difficult to rinse off.

Halogenated carboxylic acids (bromoacetic acid, chloroacetic acid) have long been favored as disinfecting agents, especially in CIP systems in the beverages industry. As with peroxyacetic acid, these products have a very wide range of effectiveness. In contrast, however, they do not lose their effectiveness when organic soiling is present. Halogenated carboxylic acids are also used in acidic media and thus formulated with phosphoric acid and/or sulfuric acid.

The law requires that cleaning and disinfecting agents must be removed from production plants in the food industry through sufficient final rinsing to the point that only technically unavoidable contamination is present. Final rinsing must occur with water that fulfills the requirements for drinking water.

Using beer as an example illustrates the negative effects of insufficient final rinsing on the beer. Quaternary ammonium compounds cause beer to cloud (protein precipitation) and destroy beer foam. They are very difficult to rinse out due to their substantivity and minimal surface tension. Aldehydes also precipitate proteins and causes beer to cloud. Products containing active chlorine can form chlorophenols through reactions with organic substances and thus considerably damage the sensory experience of beer and alcohol-free beverages. Peroxyacetic acid causes oxygen absorption and thus promotes oxidizing processes in beer. Disinfecting agent residue that does not visibly affect the beer can otherwise affect the senses unpleasantly. For this reason, the law requires that final rinsing be carried out for each agent. This is also absolutely necessary for quality reasons.

Cleaning and Disinfection

Cleaning is separation, absorption and /or removal of dirt from any surface.

Disinfection is deactivation of microorganism in order to avoid infections. We should be cleaning and disinfecting always, because the residual soil can impair the efficiency of disinfectants and the soil may shield microorganisms from the disinfectants. The factors that influence optimal cleaning and disinfection results are temperature, concentration, mechanical force and contact time.

The main components of cleaning agents are NaOH, acid, surfactants (wetting agents), enzymes, water conditioners and oxidizing agents such as chlorine or hydrogen peroxide.

In the brewing process, fat is removed by alkaline and surfactants. Protein is removed by strong alkaline. Minerals are removed by acid and carbohydrates are removed by NaOH.

Disinfection is 99.99 percent killing or deactivating of microorganism; the different kinds of disinfectants include hypochlorite, hydrogen peroxide, peracetic acid, iodophor and chlorine dioxide.

The Most Critical Aspects of Brewing: CIP (Clean-in-Place)

What is CIP?

Clean-in-place or CIP is a process by which equipment, machines, vessels, associated fittings and associated pipe work can be thoroughly cleaned without dismantling them. It can be totally automatic, semi-automatic or manual. Clean-in-place and process sterilization is one of the most critical aspects of brewing to ensure the health and safety of the consumer.

Factors Affecting Cleaning

Clean-in-place systems circulate cleaning solution in a cleaning circuit through pipe work, machines, vessel and other associated equipment. The cleaning cycle compromises different stages with water and cleaning detergents and each stage requires a certain length of time, temperature, flow, velocity and concentration of detergent to achieve an acceptable result.

Mechanical Energy (Flow in Pipes): For effective CIP, only turbulent flow is sufficient. Turbulence must be maintained in all parts of the system. Flow turbulent if >1.5 m/sec. Flow should not exceed 2.1 m/sec (waste of energy). In CIP only turbulent flow is sufficient. Turbulence must be maintained in all parts of the system.

Design Principle:

Depending on soil load on the target surface, the design of the CIP is based on one of the following:

1. *High flow rate CIP:* Deliver highly turbulent and high flow rate of detergent solution to effect good cleaning (applies to pipe work)

2. *Low flow rate but high detergent concentration CIP:* Deliver high detergent concentration solution as a low energy spray (spry ball) to fully wet the surface (applies to lightly soiled vessels)

3. *High flow rate and high detergent concentration CIP:* Deliver a high flow rate dynamic spray device (such as rotor jet) that applies to highly soiled or large diameter vessels

The Cleaning Sequence:

1. Pre-flush   : To remove gross soil
2. Caustic circulation   : To remove organic soil
3. Intermediate flush   : To flush caustic before acid cleaning
4. Acid circulation   : To remove inorganic soil
5. Sterilant circulation   : To destroy remaining microorganism
6. Final flush   : To flush out sterilant

Cleaning Detergents Used in Brewing Industry

1. Caustic Soda:

Caustic (Sodium hydroxide; NaOH) is a metallic base, predominantly ionic, containing sodium cations and hydroxide anions. It is used in many industries, mostly as a strong chemical base, in the manufacturing of pulp and paper, textiles, drinking water, soaps and detergents.

It is frequently used as a cleaning agent in the brewing industry where it is often called 'caustic'. It is added to water, heated, and then used to clean pipe works, machines, vessels, instruments and much more. It can dissolve organic soils. Surfactants must be added to the caustic solution in order to stabilize dissolved substances and thus prevent re-deposition.

Advantages:

- Excellent detergent properties when 'formulated'
- Disinfection properties, especially when used hot
- Effective at removal of protein soil
- Lends itself to automatic control by conductivity meter
- More effective than acid in high soil environment
- Cost-effective

Disadvantages:

- Degraded by CO_2, forming less effective carbonate
- Ineffective at removing inorganic scale
- Poor rinseability
- Not compatible with aluminum and other soft metals
- Activity affected by water-hardness

2. Acid:

Acid cleaning is a process employed to remove inorganic soil. Depending on the material to be cleaned, and the type of stains that are present, the cleaning acid employed may be strong or mild.

Advantages:

- Effective at removal of inorganic scale
- Not degraded by CO_2
- Not affected by water-hardness
- Lends itself to automatic control by conductivity meter
- Effective in low soil environment
- Readily rinsed

Disadvantages:

- Limited effectiveness at removing organic soil
- Limited biocidal properties
- Limited effectiveness in high soil environments
- High corrosion risk (for example, nitric acid)

3. Chelating Agents:

These are materials that can complex metal ions in a solution, thereby preventing precipitation of the insoluble salts of the metal ions. It is a substance whose molecules can form several bonds to a single metal ion. In other words, a chelating agent is a multidentate ligand. There are many sequestering agents (for example, EDTA).

4. Wetting Agents:

Wetting agents are compounds that lower the surface tension of a liquid, the interfacial tension between two liquids, or that between a liquid and a solid. Surfactants may act as detergents, wetting agents, emulsifiers, foaming agents, and dispersants. A material that can be used to reduce surface tension (wetting), the emulsification of fats or control foam, surfactants can be classified as non-ionic, anionic, cationic, amphoteric.

Cleaning Methods

Cleaning in Place is the automatic internal cleaning of production plants, such as tanks, containers, heaters, hoses and pipelines. Production plants covered with product residues are cleaned and disinfected without the need for disassembly. Cleaning and disinfecting agent solutions are sprayed onto the surfaces to be cleaned using pumps via suitable sprayers. A differentiation between 'loss' cleaning and disinfecting and 'stack' cleaning and disinfecting is made here. In both cases, the cleaning and disinfecting process is automatic.

Automation means to save time and money, facilitated working, and greater working and product safety. Cleaning in Place cleaning and disinfecting methods must ensure that all parts contacted by the foodstuff are clean. The condition here is that the production plant and CIP system must be constructed in such a way that this requirement is fulfilled.

Isolated CIP systems are required for the following production areas of a brewery:

- Brewhouse
- Pre-filtrate (fermentation tank, storage tank)
- Filtration and special systems
- Filtrate (bright beer tank, filling)

To clean and disinfect a production plant with the CIP method, cleaning and disinfecting agent solutions that are pumped through these systems in the circuit are required. These cleaning solutions must be provided by a supply system according to the consecutively occurring rinsing, cleaning and disinfecting steps.

Two method variants are available here:

- 'Non-recovery' CIP cleaning with single use of a freshly prepared cleaning solution
- 'Recovery tank' CIP cleaning with multiple reuse of prepared cleaning solutions

Non-Recovery CIP Method

The 'non-recovery' CIP cleaning system is comprised of a buffer container, pressure pump, valves, dosing equipment, heating system and a controller, and operates as follows: For pre-rinsing, water is taken from a buffer container, pumped through the system or tank to be cleaned via the pressure pump and discharged via the return line for wastewater treatment. In the next cleaning step, water is also taken from the buffer container and then circulated through the system in the circuit with added cleaning agents. Once the cleaning step is complete, the used cleaning agent is discharged. The non-recovery CIP system (loss cleaning) is more economical for small circuit volumes (less than 750 liters) than recovery tank CIP cleaning.

Advantages of non-recovery (new preparation) CIP cleaning include:

- A cleaning solution with a defined concentration is added for each rinsing step. This also allows different concentrations in different cleaning paths or cleaning programs and thus optimum adaptation to the type of soiling.

- New preparation means that a cleaning solution that is not loaded with soiling particles from previous cleanings is used.

- By discharging the cleaning solution after every rinsing step, cross-contamination is prevented.

- The plant expenditure is minimal and thus the fixed costs of cleaning are considerably reduced.

Recovery Tank CIP Cleaning Method

Recovery tank cleaning systems are generally used as centralized systems—different production areas are cleaned with various cleaning circuits from a centralized supply system. The CIP cleaning plant contains stacking containers for alkaline and acidic solutions and disinfecting agent solutions that are prepared ready for use. If possible, they are split up into application and concentration areas (Figure 12.1).

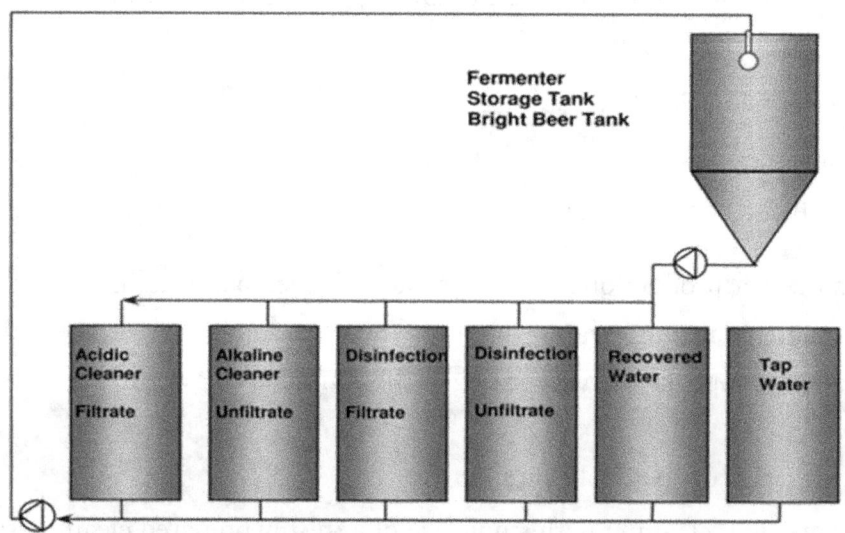

Figure 12.1 Example for a centralized CIP station.

The pre-selected concentration is maintained in each container via dosing systems comprised of a pump, mixer and conductometer. After use, these cleaning and disinfecting agent solutions are fed back into the stacking container to be used again during the next cleaning cycle. The concentration is maintained by adding the corresponding cleaning or disinfecting agent concentrate before being reused.

The CIP plant also includes tanks for fresh water pre-run, rinse water stacking, and pipe systems with valves for pre-run and return and pumps for the circuit. If necessary, heating registers are installed in the stack tanks so that they can be heated to the desired temperature directly when each rinsing step begins.

If the cleaning media are used at an increased temperature, the stack tanks should be provided with insulation. If the cleaning medium is fed to the pre-run at temperatures above 80°C, heat exchangers should be installed in the cleaning agent pre-run. Before the CIP process is carried out, the contents of the stacking containers can be heated one by one to the desired application temperature in the circuit via these heat exchangers.

The advantages of the recovery tank CIP cleaning system include:

- Lower consumption of water and heating energy for cleaning solutions thanks to reusability
- More economical utilization of cleaning solutions that must be used in higher concentrations
- Shorter cleaning times thanks to the preparation of a ready-to-use cleaning solution at the required temperature
- Shorter cleaning times thanks to the use of greater cleaning agent concentrations

Combined CIP Cleaning Method

Combined CIP cleaning systems couple the advantages of recovery tank CIP cleaning plants and non-recovery tank CIP cleaning systems, and represent state-of-the-art technology. Combined cleaning plants are different from recovery tank cleaning systems due to the following options:

- Bypassing of the stack tanks in/during the cleaning circuit
- Cleaning-specific heating during the cleaning circuit
- Option of direct dosing of cleaning agent for cleaning—specific adaptation of the required cleaning agent concentration

These technical system conditions achieve the following:

- Stack or loss cleaning can occur
- The cleaning solution needs only be heated to the required temperature relevant for the soiling
- Different cleaning agent concentrations can be used in the different cleaning circuits, since the stack tanks are only used with the concentration required for the 'simples' soiling

Permanently installed low-pressure spray balls or spray jet cleaners are currently the optimum technical solution for the spraying of cleaning solutions in CIP plants. The dimensioning, type of installation and selection of spray heads must be carried out very carefully. When doing so, the following must be taken into consideration:

- Tank shape (upright, horizontal, rectangular, cylindrical, and cylindro-conical)
- Tank dimensions
- Position and size of manholes
- Position and shape fixtures for mixers, heating coils, fluid baffles and connection sockets
- Tank materials, including seals
- Size of the outlet connection
- Type of soiling and assessment of the difficulty of its chemical detachment

The installation depth and attachment depend on sufficient flow volumes at both the top and the side tank walls. Spray balls that are too low do not achieve sufficient spray width of the spray jet in the upward direction, and spray balls that are too high deflect the spray jets so heavily due to the unfavorable angle of incidence such that the quickly descending cleaning solution could upset the entire spray jet distribution of the spray head. In the case of horizontal tanks, the installation depth is determined by the tank-filling plan of the production, or the spray balls are to be installed so low that the spray jets do not influence one another. The goal with both upright and horizontal tanks is optimum overwhelming in the heavily soiled area. The lower area is cleaned easily by the downward-flying cleaning solution. For a horizontal fermentation tank, a spray angle greater 180° is to be selected, and for a horizontal storage tank and bright beer tank, below 180°.

Selection of the type, shape, power, positioning and number of spray balls must occur based on the methods of process engineering, since the hydraulic system comprised of the sprayers and cleaning pump and the proper selection of the cleaning and disinfecting agent are the most important factors for successful cleaning and disinfecting.

The cleaning circuits are basically of two types—'open cleaning circuit' and 'closed cleaning circuit'. An open cleaning circuit is a process in which not all units (tanks, containers) to be cleaned are completely filled with cleaning fluid. This process is usually required for the pre-run and return. In the closed cleaning circuit, all pipelines and units are completely filled with water and the cleaning fluid can be lead through the system depending on the pressure. A typical cleaning program for 'open' and 'closed' cleaning circuits is specified in Tables 12.3 and 12.4.

Table 12.3: Typical CIP program for open cleaning circuit (tank)

Rinsing with fresh water/return water
Pumping down, drain
Rinsing with recovered solution
Pumping down, drain
Filling circuit
Creating circuit
Heating
Dosing (alkali)
CIRCUIT
Pumping down, recovering
Rinsing with fresh water (tap water)
Pumping down, recovering
Rinsing with fresh water (tap water)
Pumping down return water/drain
Filling circuit
Creating circuit
Dosing (acid)
CIRCUIT
Pumping down recovering/drain
Rinsing with fresh water (tap water)
Pumping down, recovering/drain
Filling circuit
Creating circuit
Dosing (disinfecting)
CIRCUIT
Pumping down, recovering
Rinsing with fresh water (tap water)
Pumping down, recovering/drain
Rinsing with fresh water (tap water)

Table 12.4: Typical CIP program for open cleaning circuit (pipeline)

[Rinsing with fresh water (tap water)/return water]
Rinsing stacked solution
Creating circuit
Heating
Dosing (alkali)
CIRCUIT
Rinsing with fresh water (tap water), recovering/drain
Creating circuit
Dosing (acid)
CIRCUIT
Rinsing with fresh water (tap water), recovering/drain
Creating circuit
Dosing (disinfecting)
CIRCUIT
Rinsing with fresh water (tap water), recovering/drain
Rinsing with fresh water (tap water)

Depending on the type and quantity of soiling in the individual areas of a brewery, there are a variety of options for cleaning and disinfecting (Table 12.5). Heavily soiled vessels in the brewhouse and pipelines are usually cleaned with an alkaline solution after pre-rinsing, and acidic cleaning occurs after the corresponding intermediate rinsing.

Table 12.5: Overview of CIP variants

Applications	Programs				
	Alkaline & acidic cleaning	Alkaline & acidic cleaning/ disinfecting	Alkaline & combined acidic cleaning/ disinfecting	Acidic cleaning & disinfecting	Combined acidic cleaning/ disinfecting bright beer tanks
	Brewhouse vessels; pipelines	FV/SV; pipelines	FV/SV; pipelines	SV/BBT	
Program steps					
Pre-rinsing	x	x	x	x	x
Alkaline cleaning (1.5–2% NaOH; 0.3% additive)	x	x	x		
Intermediate rinsing	x	x	x		
Acidic cleaning (1–1.5% acid)	x	x		x	
Combined acidic cleaning and disinfecting (1% acidic cleaner/ disinfecting)			x		x
Intermediate rinsing		x		x	
Disinfecting (0.3% peroxyacetic acid)	x	x		x	
Final rinsing	x	x	x	x	x

Fermentation tanks, storage tanks and pipelines in this area can be cleaned with an alkaline cleaner and then an acid with subsequent disinfecting or, if the contamination is less, cleaning with an alkaline cleaning step followed by a subsequent combined acid cleaning and disinfecting product. The use of alkaline cleaning agents requires a CO_2-free environment to prevent the sodium hydroxide reacting with the CO_2 and forming soda and sodium hydrogen carbonate. Soda and sodium hydrogen carbonate exhibit only minimal cleaning power. The conductivities only differ minimally, however, and are like that of sodium hydroxide. For this reason, it is necessary to monitor the concentration of sodium hydroxide via titration. The reaction between sodium hydroxide and CO_2, if it was not previously blown out with air, could lead to negative pressure (vacuum) in closed containers (cylindro-conical tanks) and thus to container damage.

Previously, bright beer tanks were cleaned with an acid and then disinfected after intermediate rinsing. The development of so-called 'one-step cleaners'—(i.e.) acidic cleaning agents containing one or more disinfecting substances—enables the cleaning and disinfecting process to be carried out in a single step. In comparison to the process described above, this process allows the elimination of two process steps. This means that, in addition to corresponding amount of time saved, water is saved due to the elimination of intermediate rinsing.

Mobile CIP systems were developed for smaller, horizontal tanks or tank systems for which upgrading with CIP spray heads was not worthwhile. A nozzle head that sprays the bundled water jets against the interior of the tank is installed on a four-wheeled base. Both the nozzles themselves (horizontal rotational axis) and the entire head (vertical rotational axis) rotate. This results in an even spray pattern over the entire interior surface of the tank. Deposits are dissolved in the cleaning fluid, and the spray jet supports this process mechanically. A pump outside the tank sucks up the provided liquid in the tank or in a separate container and feeds it to the nozzle head inside the tank through a hose lead through the manhole. The position of the carriage in the large container can be changed with the progression of the cleaning process from the outside.

From an economic and, just as important, ecological standpoint, CIP systems should be operated optimally. Cost savings are possible through the reduction of energy costs, and time can be saved in rinsing procedures, reduction of the quantities of fresh water used and, thus, the wastewater quantity, and through the proper usage concentration of the cleaning and disinfecting products. Analysis tools that can measure the temperature, pressure, conductivity value, volume flow and valve setting parameters during a cleaning and disinfecting step are available today. They are then displayed graphically after the analysis.

After the analysis and interpretation of the measurement results (actual state), it can be determined whether the cleaning and disinfecting programs are set optimally or whether there is potential for savings. Rinsing with fresh water in a brewery was carried out for far too long after a cleaning step. In this example, you can see that a reduction in the fresh water rinsing time from ten to five minutes is sufficient. This corresponds to a savings of 1.6 m³ of water per cleaning step. Accordingly, the amount of fresh water saved during one year's worth of cleaning would be 2,880 m³. Considering the costs of tap water and wastewater, huge savings could be achieved with this plant alone.

Automation and Monitoring

Complex CIP systems are ideally controlled automatically, the advantages being:
- A program can be designed and set when the plant is commissioned to maximize cleaning, and this will be consistently adhered to

- Detergent and sterilant strengths can be optimized
- A cleaning cycle can run unsupervised
- Automated recording of cycle times, detergent strengths and temperatures by the monitoring equipment is available
- Cleaning can be held up if a problem is detected
- Sensors can detect detergent/sterilant strength on the return line and direct the return to tank or drain saving chemical costs

In the CIP system illustrated below, the following items are automatically controlled:

- Inlet and outlet valves of detergent/sterilant tanks
- Delivery pump
- Rinse water and drain valves
- Detergent/sterilant strength detection and tank top-up from bulk supplies
- Detergent/sterilant strength detection on the return line

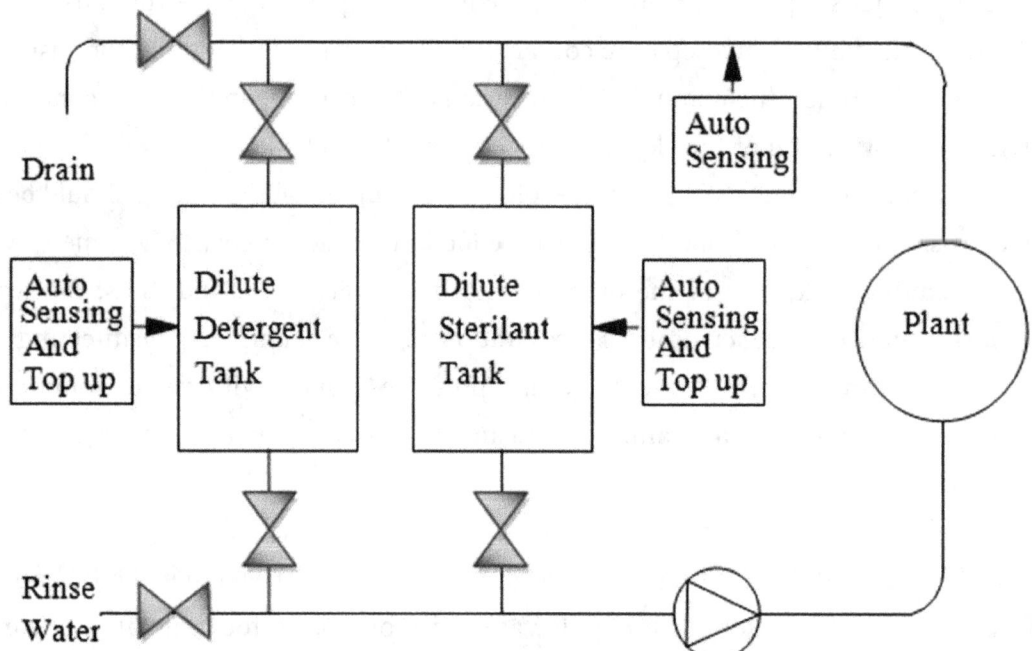

Figure 12.2 Example for an automated CIP station

Material Compatibility

The most common material used in the manufacture of tanks, equipment and pipelines is chrome-nickel-steel. Aluminum, copper and standard steel with linings (for tanks) are also still used, albeit to a lesser degree. Chrome-nickel-steel is characterized by its very good chemical resistance to all alkaline and acidic cleaning agents in common concentrations, with the exception of cleaning agents containing sulfuric acid. Disinfecting agents containing hypochlorite (active chlorine) and water with a heavy concentration of chloride ions can lead to pitting, depending on the concentration, temperature and contact duration. When using cleaning agents containing active chlorine, it is absolutely imperative that they not be allowed to mix with acidic cleaning solutions (due to the formation of chlorine gas).

Aluminum is generally attacked by all acidic and alkaline cleaning agents. However, cleaning agents containing sulfuric acid and nitric acid create chemical-safe passive coatings under certain circumstances. In the case of cleaning agents containing phosphoric acid, the amount of weight lost can be categorized as consistent. Alkaline cleaning agents are especially harmful to aluminum unless special corrosion inhibitors are contained in these cleaning agents.

System parts made of copper are to be viewed critically from a corrosion standpoint if oxidizing cleaning solutions are to be used. Oxygen released in the cleaning solutions also accelerates the harm done to the material. In CIP circuits that contain system parts made of copper or copper alloys, only cleaning agents with special corrosion inhibitors should be used for this reason; oxidizing cleaning agents, such as cleaning agents containing nitric acid, chlorine or active oxygen lead to increased surface attack or discoloration.

Coated tanks and equipment in particular are designated as being highly susceptible to corrosion from a cleaning standpoint if damaged linings (hair fractures) come into contact with acidic cleaning solutions. The resistance of lining materials to cleaning agents must be obtained from the respective manufacturer, due to their differing chemical compositions (epoxy resins, phenolic resins, epoxy-phenolic resins, glass enamels). In general, glass enamel and epoxy resin coatings, phenolic resin coatings and epoxy-phenolic resin coatings exhibit good resistance to acidic cleaning agents at room temperature. Alkaline cleaners, on the other hand, should not be used above a specified upper temperature limit. Essentially, they cannot be used with phenolic resin and epoxy-phenol resin linings. Oxidizing cleaning agents (active chlorine, active oxygen) may only be used after preliminary testing or agreement with the manufacturer of the tank lining.

If sealing materials come into contact with food, they must exhibit certain levels of cleanliness. Since cleaning and disinfecting can almost never be carried out with only a single chemical type, the optimum material must be selected from the available range. In the case of seals, consideration of the chemical resistance and the swelling behavior are often decisive for the proper selection of a suitable material. In addition to the multitude of factors affecting the seal, other influential values must be taken into account (for example, the temperature, resistance to grease, pressure, speed, surface composition of metallic parts and the material type of the machine parts to be sealed). Compatibility tests should also be carried out in this case before making a decision on the sealing materials to be used.

Cleaning Glass Bottles

Industrial bottle cleaning is a complex, engineered process that is subject to many dependencies. Cleaned bottles in the beverages industry are subject to the following requirements:

- Freedom from typical empty bottle soiling, such as mold, dust and dried-on beverage residues
- Freedom from any bottle covering (labels)
- Freedom from beverage-spoiling germs
- Clear, shiny appearance
- No odors present
- No chemicals present
- A bottle temperature suitable for filling

Bottle cleaning in the brewery and beverages industries has always been important. Several types of machines with different structures and functions have been developed over time. Modern bottle cleaning machines operate with the combined soak-rinse process. Here, the soaking zones are used for loosening the soiling and spraying removes the loosened soiling so that the cleaning solution can be optimally effective on the remaining soiling. Essentially, a differentiation is made between single-end and double-end machines. Both machine types have specific advantages and disadvantages, but these have no bearing on the effectiveness of cleaning. While German breweries predominantly use single-end machines, double-end machines are often used for bottle cleaning in breweries in other countries.

Bottle Cleaning Machine

A modern bottle cleaning machine is constructed based on the following treatment zones:

- Residue emptying
- Pre-soak
- Caustic soak zone (several possible)
- Caustic spray (several possible)
- Intermediate spray zone
- Hot water zone (several possible)
- Cold water zone
- Fresh water zone

The individual zones in the bottle cleaning machine carry out the following tasks:

Residual Draining

By rotating the bottles, any liquid found within can run out and be removed from the machine, and thus does not affect downstream zones.

Pre-soak with Pre-spray

In the pre-soaking zone, the bottles are filled with water (drained, heated water from the intermediate spray zone) and then emptied again. This removes most loose, clinging contamination and beverage residues from the bottles so that as little soiling as possible ends up in the main caustic soaker bath (which results to longer useful lye life and therefore less alkali consumption). In addition, the CO_2 gas found in the bottles is removed from the bottles and partially neutralized via the preheating water. Pre-spraying and vapor haul-off ensure that no CO_2 or only minimal CO_2 ends up in the alkaline area, which would neutralize the cleaning lye (sodium carbonate formation).

In addition, the bottles are slowly heated by the draining intermediate spray water. The pre-caustic is moved toward the post-caustic via a heat exchanger, whereby a large amount of the heat from the post-caustic can be recuperated. This heat exchanger saves lots of energy in heating up bottles.

Main Caustic Soaker Bath and Caustic Spray

Caustic soaker baths are the most important zones of the bottle cleaning machine. These baths are responsible for the main cleaning due to their temperature being the highest (beer residue is broken up, greasy contamination is saponified and can be emulsified, and protein substances are denatured and detached from the bottle surfaces). Labels are detached from bottle surfaces through the combination of heat, sodium hydroxide, additive (concentrate) and contact duration.

The number of main caustic soaker baths is determined based on the degree of bottle soiling. In the case of slightly soiled bottles, an alkaline immersion bath with an approximately six minutes immersion time is sufficient to properly clean the bottles.

In the case of bottles with heavy contamination (such as mold deposits and dried-on soiling), at least two caustic baths with a spraying session in between them must be used to properly clean the bottles for an immersion time of twelve to fourteen minutes. Caustic-spray zones and the label-discharge systems are located at the end of the soaker baths. Loose clinging contamination is rinsed off here, and labels located between the bottle and bottle cell are rinsed out of the cells and carried away via the label-discharge system.

A very good cleaning result can be achieved within a short amount of time through the combination of a caustic soaker zone and caustic spray.

Intermediate Spray, Hot and Cold Water Zones

These zones are usually designated water zones of the bottle cleaning machine. These water zones are used to rinse off alkalis and other residues and to cool the bottles at the same time. Fresh water is usually relatively cold. This ensures that the bottles exit the cleaning machine at low temperatures.

The cascaded connection of the water zones causes the drained and slightly heated fresh water to end up in the cold water zone, then in the hot water zones and subsequently in the intermediate spraying zone. At this point, the water has a considerably high alkali content and temperature and runs into the pre-soak.

Use of Chemicals in the Bottle Washing Machine

Depending on the type of contamination and the consistency of the water used (hardness), different chemicals are used in the different zones of the bottle washing machine. Sodium hydroxide is usually used as the base chemical for cleaning in the main soaker bath. In addition to the known properties of sodium hydroxide, other negative properties are relevant in this area:

- The high surface tension means that penetration into labels is reduced (label removal takes too long)
- The labels can become frayed
- The bottle material (glass and PET) can be damaged
- Disturbing foam via saponified contaminants

The alkaline cleaning solutions are comprised of a fully prepared cleaner or from sodium hydroxide and a caustic soda-free additive (concentrate). As there are different types of contamination to be removed (beverage residues, label glue, label paper, printing ink and so on), there can be no 'all-round cleaner'. The additive for cleaning refillable bottles must be selected based on the plant and cleaner types. There are additives that are geared toward, for example, water-hardness stabilization, mold removal, rust ring removal, high complexing capacity and reduced scuffing to glass by sodium hydroxide. The contents of the additives are listed in Table 12.1.

Labels are to be discharged with as little decomposition as possible. If the labels are frayed due to poor paper quality or an excessive concentration of the alkali, undesired foaming may occur. In this case, a separate defoamer must be added in doses to prevent over-foaming of the lye and to ensure trouble-free discharge of labels.

In the water zones, the carried caustic soda is removed from the hot surface of the bottle. Increasing the temperature or pH value can cause hard water deposits. This can be prevented by the use of acidic products that contain sequestering and dispersing agents in addition to the acid. These products operate in the sub-stoichiometric range and prevent the formation of hardness-mineral crystal grid structure. This means that they do not precipitate as solid, hard deposits, but rather as fine amorphous particles that suspend well and can be rinsed out.

Complexing agents are not used for economic reasons. They would prevent precipitation completely, but would have to be added in doses of higher concentrations, since they form complexes with the metal ions of the minerals in the water causing hardness 1: 1.

Cleaning PET Bottles

Bottles made of PET material (polyethylene terephthalate) are cleaned using the same system as glass bottles. Chemically, PET is polyester (plastic) with temperature and chemical resistance properties that must be taken into consideration. One particular problem in comparison to glass is the temperature resistance of the material. At temperatures over 60°C, the material shrinks. Therefore, the cleaning process may not exceed a maximum temperature of 60°C.

If PET material (polyester) is exposed to pure sodium hydroxide at greater concentrations, stress cracking and hazing occurs. As with the cleaning of glass surfaces, additives for lyes that support the cleaning process must also be used for PET bottle cleaning. Most additives for caustic solutions used with glass cannot be used, however, as they contain substances that harm the bottle material. Appropriate additives that support cleaning and effectively prevent the occurrence of stress cracking and hazing were developed for use in the cleaning of PET bottles.

If the defoaming action of the lye additive be insufficient, a suitable defoamer must be used separately, as with glass bottle cleaning. It must also protect the PET material from hazing and stress cracking.

Foam Cleaning

The cleaning of non-closed systems, such as the external surfaces of tanks, pipelines, machines, open containers, walls and ceilings, usually occurs via low-pressure foam cleaning. In the case of the listed objects to be cleaned, they are usually large surface areas that may be only poorly or completely inaccessible. High tanks, areas under tanks, areas between pipes or very irregular, angular surfaces of machines (fillers) often pose problems when it comes to cleaning. With foam cleaning, relatively long contact times between contaminants and cleaning agents are achieved with low liquid consumption. This process is used especially for bulky objects and on vertical surfaces. The long soaking times are especially advantageous when dried-on and burned-in contaminants must be soaked and swelled before they can be sprayed off with water.

Foam cleaning has the following advantages:

- The detection of all surfaces to be cleaned is unambiguous, since untreated areas of foamed areas are easy to distinguish visually
- In comparison to previously executed manual cleaning with brushes, scrubbers and cloths, the risk of infection with foam cleaning is reduced considerably
- In comparison to manual processes, the capacity for treating large areas is greatly increased
- The cleaning personnel do not come into direct contact with the cleaning solution

This type of surface cleaning of large-surfaced objects is carried out with systems working with pressure up to 40 kg/sq. cm. Here, the water inlet pressure is increased by approximately 20 kg/

sq. cm in the units. To generate the necessary foam, a partial current of the high-pressure water is deflected via an injector system and the previously set quantity of cleaning agent is drawn into the water stream via under-pressure there. The air required for the foam generation reaches the water/cleaner mixture via a dosing unit and creates the desired foam in the unit, which can then be applied via a hose no longer than 25 m, a gun or a special nozzle. Foam cleaning is usually combined with spraying procedures for pre- and post-cleaning.

Pre-cleaning often occurs at a slightly increased temperature (40–50°C), mainly to increase the effectiveness of the subsequent foam by heating the surfaces. The temperatures specified for foaming the cleaning solution vary between room temperature and 85–90°C. The soaking time of the foam on the contamination is usually between ten and twenty minutes, whereby stubborn contamination can extend the contact time to forty minutes if the foam is sufficiently stable.

The subsequent high pressure cleaning usually occurs with hot fresh water at approximately 50°C. In many cases, it is recommended to use subsequent spray disinfecting (filler).

Work Safety and Environmental Protection

Cleaning and disinfecting agents are more or less dangerous chemicals. To sufficiently protect the health of the personnel handling these substances, the information on the material safety data sheet must be observed. In general, it should be mandatory to wear safety goggles and the recommended protective clothing when handling chemicals.

The product information sheet on cleaning and disinfecting agents should also be read thoroughly and observed. Recommended usage concentrations and temperature specifications are to be complied with, which also protects the environment.

Used cleaning agent solutions are often heavily alkaline (pH >11) and must be neutralized to the legally prescribed pH value before being released into the public sewerage system.

Validation of CIP

It is essential that the procedures be followed to ensure that CIP processes are carried out correctly. Two types of validation procedure are used. Firstly, checks must be made to confirm that the conditions employed during the cleaning process were within the predetermined specification. Secondly, the cleanliness of the plant must be assessed against predetermined standards.

Cleaning in place checks include cycle times, temperatures and the strengths of cleaning chemicals. Assuring that the correct concentration of caustic soda-based detergent is used is worthy of special comment. Commonly, the strengths of solutions of caustic soda are assessed by measurement of conductivity. Such readings can be misleading since conversion of sodium hydroxide to sodium bicarbonate following exposure to carbon dioxide does not produce a change in this parameter. Preferably, the concentration of caustic soda solutions should be checked by off-line titration.

The cleanliness of plant after CIP can be checked using adenosine triphosphate (ATP) bioluminescence. Where possible, this should be supplemented with visual checks to ensure that spray balls are functioning correctly and no shadow areas exist. Recently the use of a video camera, termed the 'topscan', mounted in the top of cylindro-conical fermenters has been recommended for examining vessel cleanliness. Validation of CIP is essential when new plant is commissioned. Since CIP is a costly and time-consuming process it is necessary to employ conditions that provide the desired level of cleaning at the lowest cost.

13

Quality Control

QUALITY CONTROL IS the total collection of all means and activities by which the standardized properties of a designed product are precisely and consistently maintained throughout production, marketing and consumers' use.

The objectives of quality control are to insure the uninterrupted integrity of product and process, to avoid quality jeopardy, and to avoid recourse to 'corrective action'.

Consumer acceptance, based on quality, depends on the following:

1. Recognition and interpretation of the preferences existing in a large population
2. The recognition and application of the concept that consumer preferences are paramount to product design
3. A product that accurately incorporates consumer preferences
4. A production quality control regime that strives for uncompromising excellence in producing the designed product
5. Consistent conformance to the total product design and Good Manufacturing Practices (GMP) from conception through final packaging
6. A corporate image of excellence earned through the continuous and consistent distribution of quality products

General Brewery Functions

Regardless of the brewery capacity, seven functions must be accommodated with respect to administrative and primary technical responsibilities are:

1. Production operations
2. Process control
3. Quality control
4. Quality assurance

5. Research and development

6. Procurement

7. Technical advisory committee

These functions are interrelated and, as a unit, are bound to all phases of production and distribution.

A. Production Operations

Quality of product and process begins with the production operators—their performance primarily determines the excellence of a quality program. Production operations responsibilities are:

- Assessment of performance
- Conformance to specifications
- Guidance
- Documentation
- Institution of corrective action

The principal positive quality action is operator performance. To be successful, the brewer must place the highest emphasis of quality control on the training and performance of the operator and the equipment.

B. Process Control

Process control is an 'on-line' quality control tool operating in-line function with production. A program of routine technical surveillance is included in the assignments of the group. Close technical liaison between this group and quality control provides analytical procedures, reagents, instrumentation, counsel, and an extension of services to production. An added responsibility of this group to production personnel is that of 'on-site' technical interpretation of observations and analytical results.

C. Quality Control

The quality control department is the central technical group responsible for management of the quality program. All specifications for materials, processes, and the product itself must be established by this group. It is quality control's responsibility to be aware of, at all times, the quality profile of the materials, processes and products, and it must honestly, clearly and promptly convey this information to the head of the brewery.

In summary, quality control is neither a faultfinder nor a 'witch hunter'—it is the navigator to product excellence.

D. Quality Assurance

The quality assurance department is the quality conscience of the brewery. It is the source of all materials, process and production specifications, guidelines, process procedures, audits and quality appraisals. It must work cooperatively in technical liaison with all departments. It must also interface with regulatory agencies, materials and equipment suppliers, distributors, retailers and customers. Necessary laboratory and analytical services are provided by the quality control department or, in special cases, by consulting laboratories.

The more important services of the quality assurance department are collect, assemble, and keep current complete and accurate, detailed information on the following:

- Materials
- Formulations
- Process procedures
- Process yields
- Product composition
- Additional information such as description of the process and products

E. Research & Development

The role of research and development is to be cognizant of the state-of-the-art technology and to extend proprietary intelligence to improved quality, operations, and effectiveness. Production and quality control represent strict and constant adherence to a rigidly defined system. Research and development represent change and flexibility. Despite an extreme difference in philosophies, both functions are useful and should provide valuable services to each other.

The research and development division is to be administered separately from the production and quality control functions to an extent that insures their absolute protection from any research influences or duress in matters of their operations, decisions, or practices.

F. Procurement

The procurement department executes and controls all purchases made by the brewery; it insures the value received and delivery of the goods all according to *good business practices*. The first requirement of a purchase is the identification and precise description of the commodity—this establishes the quality of the purchased items. It is critical that this level of quality be consistent with the level of the quality of the system to which it is applied. The procurement officer should be a well-informed, intimate participant of the quality team. Procurement cannot be expected to meet its responsibilities without the counsel and full support of the quality control/quality assurance departments. However, acceptance of all process-related materials is solely the prerogative of quality control. The provision of high quality brewing materials is the foundation for 'building' quality into a product.

G. Technical Advisory Committee

A technical advisory committee counsels management on complex technical issues involving more than a single area of professional expertise. The team usually includes the heads of quality control, quality assurance, production, engineering, and research. The respective section-heads are included in the deliberations upon specific request.

Product Definitions

No product can be manufactured without knowing an established identity and description. It follows that no quality beer can be manufactured without a design that includes its comprehensive description, definition and specifications. The preparation of precise, detailed product descriptions and specifications for each product is a fundamental necessity before any production is to begin. If it is found that an operating brewery does not have such documents, they should be promptly prepared. It is only through precise knowledge of goals, direction and guidelines that excellence of quality in a product can be attained.

The development of a new product or the redesign of an existing one begins with its description. Then follows, in preferred succession throughout the process:

- Formulation
- Process
- Process procedure
- Materials specifications
- Product specifications

This is tedious task but absolutely necessary for the foundation of a successful quality control program. This must be done honestly and realistically, describing exactly the degree of perfection to which the management commits itself.

Quality Control in the Brewery

For quality control of beer, the quantitative determinations of the followings parameters are essential.

1. Wort production
Process quality control parameters for wort production should be extract, color, bitterness (BU), pH and dissolved oxygen.

2. Pitching
Process quality control parameters for pitching yeast should be cell count, dead cell and yeast pitching rate. (Wort bacteria, wild yeast and lactic acid bacteria infection may be applied)

3. Fermentation

Process quality control parameters for beer fermentation should be apparent extract (˚Plato), temperature, top pressure, cell count, dead cell, growth rate, pH, VDK and flavor control (wort bacteria, wild yeast and lactic acid bacteria infection may be applied).

4. Maturation

Process quality control parameters for beer maturation should be temperature, top pressure, original extract (˚Plato), alcohol, apparent extract, bitterness, pH, VDK and flavor control (wild yeast and lactic acid bacteria infection may be applied).

5. Stabilization

Process quality control parameters for beer stabilization should be CO_2 content, temperature, top pressure, dissolved oxygen, polyphenols content, anthocyanogens contents, colloidal stability (wild yeast and lactic acid bacteria infection may be applied).

6. Filtration

Process quality control parameters for filtration should be turbidity (haze), dissolved oxygen, CO_2 content, original extract or alcohol (wild yeast and lactic acid bacteria infection may be applied).

7. Bright Beer handling

Process quality control parameters for bright beer should be original extract, alcohol, apparent and real extract, RDF, pH, color, turbidity (haze), CO_2 content, dissolved oxygen, SO_2 content, head retention (foam stability), bitterness, VDK and flavor control (wild yeast and lactic acid bacteria infection may be applied).

8. Packaging

Process quality control parameters for packaging should be CO_2, total oxygen content and air in headspace.

Microbiological Quality Assurance and Quality Control

The brewery microbiologist should be clearly involved in the entire beer production operation. Followings are the key areas of consideration:

- Inspection of incoming raw materials, including water
- Examination of brewing equipment and the brewing environment
- Microbiological surveillance over the product at all stages
- Evaluation of cleaning materials and cleaning/sterilizing procedures
- Training of laboratory staff and liaison with brewing, engineering, and management members

Quality control (QC) and quality assurance (QA) in the context of brewing distinguish the act of determining the current or very recent microbiological status of the plant and products (QC) from the actions that are put into place to ensure a quality standard (QA). The results of QC are used to provide QA—for example, dirty surfaces in fermenters must be cleaned or poor viability yeast discarded. It is important that the process is not entirely reactive; however, a proactive approach should be adopted to prevent faults occurring. There are essentially two approaches to microbiological testing: Traditional methods rely on cultivation of yeasts and bacteria in appropriate media followed by identification of the offending organism, if necessary, and modern rapid approaches have a minimal reliance on prior cultivation. The move toward rapid methods for microbial detection and identification has been driven by technological developments and also by changes in the industry. Some key motivating factors are:

- Growing market volumes for non-pasteurized beer in cans and bottles
- More low- and non-alcoholic beers
- Increasing variety of flavored sweetened alcopop-type beverages
- Tightened government regulations

Minimizing the time needed to detect a spoilage agent can lead to significant savings through reduced product recalls, extension of shelf life, and consistency of product quality and flavor.

The decision to adopt traditional or rapid methods is to an extent dependent on the critical control point (CCP) under review. For example, processes in the brewhouse, fermenting hall and storage cellar generally require at least five days—so, traditional methods are appropriate to provide the green light for the next step. However, other stages such as filtration, bright beer cellar tanks, CIP, and water services are more constrained by time and rapid tests can provide the necessary information for optimization of the process.

Setting Microbiological Standards and Sampling

Hazard analysis of critical control points (HACCP) has become an essential feature of QC in the food and drink industries, including brewing. This involves the systematic assessment of all the steps involved in the brewing process and identification of those steps essential to the hygienic quality of the product. In a large, complicated plant, it would be advisable to reduce the operations to a series of connected subroutines.

Many of the CCPs are not microbiological (the principal microbiological CCPs are shown in Figure 13.1). At these stages it is essential to monitor for microbiological hazards by using either traditional or rapid methods. First, it is necessary to set microbiological standards for each point. What is the maximum allowable level of contamination and by what organisms? Some suggested levels of sensitivity for detection are given in Table 13.1, but these are guidelines only and it is important to establishing own definitive criteria. For example, it may be permissible to use pitching yeast with limited bacterial contamination by *O. proteus*, but yeast contaminated with *Pediococcus* should be discarded and the source of contamination determined. Some organizations adopt a 'green, amber, red' approach—green flags adherence to microbiological standards (no action needed); amber indicates minor microbiological concern, the brewing process can continue but some microbiological contamination has occurred, and should be investigated; red indicates failure to meet a microbiological standard and production staff must be informed so that corrective action can be taken. It is valuable to prepare trend graphs showing the microbiological status over time for the various stages and products. A trend showing a gradually worsening microbiological situation can give early warning of a problem.

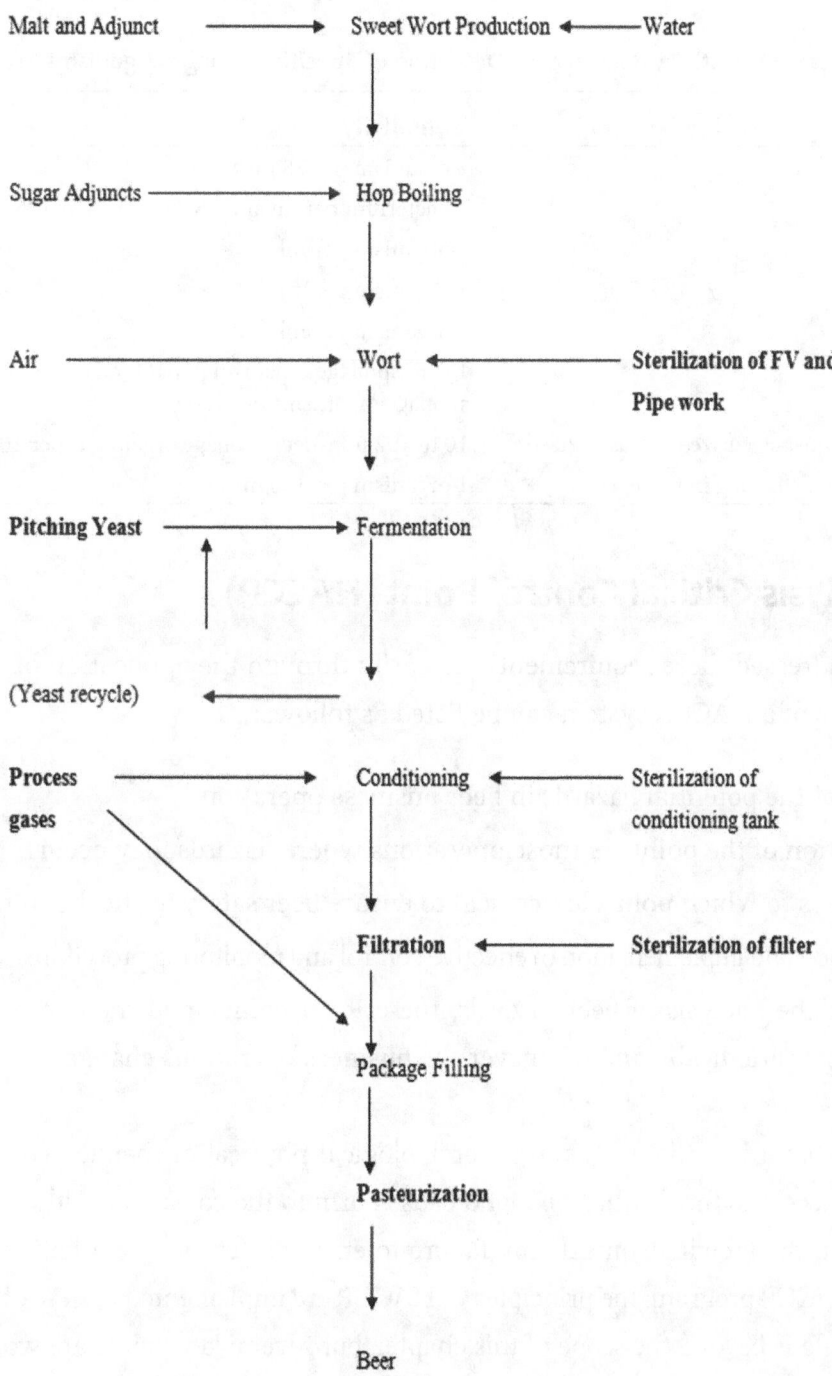

Figure 13.1: Flow diagram of beer production showing critical control points (CCPs) for microbiological testing in bold

Sampling at the CCPs requires careful consideration so that the results of the microbiological tests are statistically sufficient. Typical samples are liquids—for example, samples of final water rinses from CIP operations or water rinses of containers. Such samples should be of sufficient scope to provide for assurance that the microbiological standard has been achieved. For example, when examining beer, a sufficient quantity should be filtered or forced to allow for detection of contaminants at the required level, usually 100 to 1000 ml for testing of packaged beer depending on the size of the container and the sensitivity required (Table 13.1).

Table 13.1: Suggested Sensitivity Required for Detection of Specific Spoilage Organisms in Brewery Samples

Samples	Sensitivity
Cold aerated wort	1 organism per 25 ml
Pitching yeast	1 bacterium per ml and 1 wild yeast per 106 culture yeast
Fermenting wort	1 organism per ml
Tank bottoms	1 organism per ml
Beer in storage	1 organism per ml
Filtered beer	1 beer spoilage organism per 100 ml or 10 to 102 non-beer spoilage organisms per 100 ml
Packaged beer (non-pasteurized or pasteurized)	10 to 102 non-beer spoilage organisms per 100 ml
Rinse water (end of cleaning in place)	1 organism per 100 ml

Hazard Analysis Critical Control Point (HACCP)

Brewers have addressed these requirements primarily through the application of HACCP methods. The key elements of a HACCP system can be listed as follows:

1. Analysis of the potential hazards in beer business operation
2. Identification of the points in those operations where hazards may occur
3. Decisions as to which points are critical to ensure beer safety (critical points)
4. Identification and implementation of effective control and monitoring procedures at the critical points
5. Review of the analysis of beer hazards, the critical points, and the control and monitoring procedures, periodically and whenever the business operations change

In HACCP terms, hazards are defined as any microbiological, physical, or chemical contaminant that may potentially gain access to the finished beer and cause harm to the consumer. Although hazards clearly relate to safety, in the brewing context, any failure to ensure safety is also a failure in a QMS. When carrying out a HACCP program, the principles of HACCP are implemented review of the application of HACCP in brewing is beyond the scope of this chapter, but several key points are worth emphasizing:

1. Prepare a Flow Diagram: The purpose of the flow diagram is to provide a detailed description of the process to help the HACCP team carry out the hazard analysis. The flow diagram is

essential to the HACCP team when identifying hazards in the process. The flow diagram should be an activities diagram, showing each process step in the order in which it is carried out, including rework routes. All material additions and services should be shown in the diagram.

2. Identify the CCPs: A critical control point is a step or procedure in the brewing process where control is essential to prevent, eliminate, or reduce a hazard to an acceptable level. The World Health Organization (WHO) recommends that CCPs should be determined using the HACCP decision tree.

Usually, HACCP has been applied where processes present hazards to health from either physical or chemical agents. Microbiology has not usually been included except where there might be a health risk from pathogenic organisms. It is beginning to be used as a method of microbiological quality assurance where the risk is limited to product wholesomeness.

There are several elements to HACCP analysis. The starting point is to draw up a detailed flow diagram of the process under consideration. Once constructed, the process flow diagram must be verified to ensure that all relevant steps have been included. All HACCP analyses contain seven parts (Fig. 13.2).

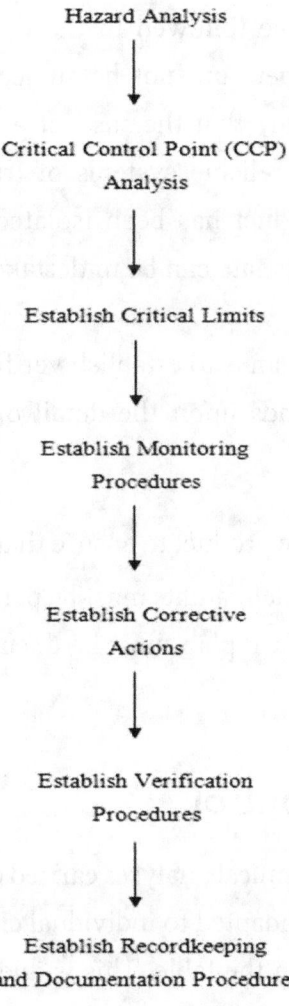

Fig. 13.2: Elements of HACCP Analysis

- In the first part, each step in the process is assessed and its inherent risks are identified.

- In the second step, the identified risks are graded to identify those that are critical control points (CCPs). A CCP is a process step that, if not under proper control, has the potential to cause injury or illness to consumers. For microbiological control of the brewing process, the concept of the CCP is widened to include a potential to result in the sale of product that is not wholesome.

- The third part of the analysis is to set critical limits for each CCP. For example, in the case of a pasteurized beer, the critical limits would be the time and temperatures needed to ensure that the product is rendered microbiologically stable.

- The fourth step is to establish monitoring procedures to ensure that the critical limits are adhered to and the CCP is under control. In the case of the pasteurized product this would be a permanent record of the times and temperatures to which all batches of product had been exposed.

- The fifth part of the analysis establishes corrective actions should a CCP be found to be out of control. To continue the example of the pasteurized product, this would include procedures to segregate product, which it was suspected might not have received the specified heat treatment. The procedures to be followed in these circumstances must be detailed in the HACCP plan. They must happen and not be subject to discussion. In the example cited, the first priority would be ensure that the suspect product could not be sent to the market. The absolute requirement for reliable systems of traceability and labeling can be readily appreciated. Once suspect product has been isolated a more leisurely examination of the problem and consideration of its fate can be undertaken.

The penultimate step in the HACCP plan is to establish verification procedures. These are of several types and their precise nature depends upon the detail of the process. They must include two elements.

1. A day-to-day examination of the product to ensure that it meets pre-established specifications.

2. Regular and preferably independent audits must be performed to guarantee the integrity of the process. Finally, the entire HACCP plan must be documented and a system of recordkeeping set up.

Analyses in Daily Quality Control

Table 13.2 shows the most frequent chemical analyses carried out at the different stages of the process. This is only an example and has to be adapted to individual circumstances. Not all analyses described in the first part of the chapter occur in the table. This is due to the fact that some analyses like the determination of oxalic acid, hop oil or in finding the cause of problems like filtration difficulties or gushing, but have not found their way into day-to-day quality control.

Table 13.2: Example of a control scheme for breweries (day-to-day control)

Sample	Analyses	Frequency
Raw materials		
Malt	Friabilimeter	Every delivery
	Sieving test (grading)	Every delivery
	Hand assessment	Every delivery
	Water content	Each supplier every 3 months
	Congress mash	Each supplier every 3 months
	Extract	
	Wort color	
	Color of boiled wort	
	Viscosity	
	pH	
	Final degree of fermentation	
	β-glucan	
	DMS precursor	
	Homogeneity and modification	Every supplier every 3 months
	Contamination	Every supplier at least once per year/every 3,000 tons of malt consumption
	Mycotoxins	Every supplier at least once every 6 months
Hops	α-acid content	Each delivery
Water		
Water intake	Smell and taste	Once per week
	Conductivity	Once per week
	Turbidity	Once per week
	pH	Once per week
	Complete analyses:	Once per year or according to referring legislation
	All ions	
	Heavy metals	
	Contaminants:	
	Pesticides, etc.	
	Trihalomethanes (THMs)	
	Further organochlorides	
	Polycyclic aromatic	
	Hydrocarbons (PAHs)	
	Benzene	
	Acrylamide	
	Epichlorhydrin	
	Vinyl chloride	
	Color	
	Total organic carbon	
Brewing water	Smell and taste	Once per week
	Conductivity	Once per week

(continued)

Sample	Analyses	Frequency
	pH	Once per week
	Total hardness	Once per week
	Alkalinity (m value)	Once per week
	Residual alkalinity	Once per week
Service water	Disinfectant, e.g. CO2	Once per week
	pH	Once per week
	Total hardness	Once per week
Boiler feedwater	Total hardness	Daily
	Conductivity	Daily
Auxiliary Material/Aids		
Kieselguhr	Odor	Every delivery
Polyvinylpoly-pyrrolidone (PVPP)	Odor	Every delivery
Acids	Concentration	Once every 3 months
	Purity	Once every 3 months
Caustics	Concentration	Once every 3 months
	Purity	Once every 3 months
Disinfectants	Concentration	Once every 3 months
	Purity	Once every 3 months
Process control		
Grist	Visual evaluation	Every charge
Mash	Saccharification (iodine test)	Every brew
Lauter wort	Extract	Every brew
	Turbidity	Every brew
	Odor and taste	Every brew
Spent grains	Soluble and digestable extract	Once every 3 months
Cast wort	Extract	Every brew
	Final degree of fermentation	Every month
	pH	Every month
	Color	Every month
	Bitter units	Every month
	TBI	Every month
	Viscosity	Every month
	β-glucan	Every month
	Nitrogenous compounds FAN	every month
Green beer	Original gravity, extract, alcohol, degree of fermentation	Every fermentation
	Vicinal diketones	Every fermentation
Unfiltered beer	Original gravity, extract, alcohol, degree of fermentation	Every tank, 1 day before filtration
	pH	Every tank, 1 day before filtration
	CO_2	Every tank, 1 day before filtration

Sample	Analyses	Frequency
	Turbidity	Every tank, 1 day before filtration
	Sensory analysis	Every tank, 1 day before filtration
Filtered beer	Oxygen (in-line)	Every filtration charge
	CO_2 (in-line)	Every filtration charge
	Turbidity (in-line)	Every filtration charge
	Sensory analysis	Every filtration charge
Beer in Bright Beer Tank (BBT)	Oxygen	Every tank
	CO_2	Every tank
	Sensory analysis	Every tank
	Original gravity, extract, alcohol, degree of fermentation	Every tank
	pH	Every tank
	Color	Every tank
	Bitter units	Every tank
Filling	Original gravity, extract, alcohol, degree of fermentation	Every charge
	Color	Every charge
Filled Beer	Original gravity, extract, alcohol, degree of fermentation	Every charge
	pH	Every charge
	Color	Every charge
	Bitter units	Every charge
	CO_2	Every charge
	Foam	Every charge
	Turbidity	Every charge
	Sulfur dioxide	Every product every 3 months
	Polyphenols and anthocyanogens	Every product every 3 months
	Higher alcohols	Every product every 3 months
	Steam evaporable fatty acids	Every product once per year
	Polyphenols and anthocyanogens	Every product every 3 months
	Higher alcohols	Every product every 3 months
	Steam evaporable fatty acids	Every product once per year

Criteria for the decision to carry out the analysis in-house or by an external lab are:

- The necessity of the analysis (legislation, demands from the retailers, quality)

- The time span in which the results have to be available

- Availability of in-line measurements

- One's own laboratory capacity and qualification

When using in-line measurements, it has to be kept in mind that the availability of qualified personnel and of a reference measuring system in the lab is absolutely essential. Therefore, in-line measurements may help the laboratory in its daily work, but will never be able to substitute it as a whole.

14

Sanitation and Pest Control

Introduction

The modern brewery could be host to microbiological problems, such as stray microorganisms, and to macrobiological problems, such as insects, rodents, and birds. We would never eat from a dirty, cracked plate because we know that bacteria could live in the cracks in the plate and reproduce on the small amount of food on it. By the same token, most insect and rodent control depends on similar cleaning and repair to remove the sources of food and shelter for these pests. Sanitation and pest control must go hand-in-hand.

The warm, moist atmosphere of a brewery and the use of grains as ingredients contribute to the attractiveness of the brewery to pests. It is necessary to examine what can be changed to make the plant less attractive to pests. One also needs to have well-sealed buildings and physical barriers to deny entrance to these pests. These points will be discussed in detail later—first, it is important to first look at the types of pests that may be encountered in a brewery.

Types of Pests Encountered

Pests can be classified in a number of ways but they will be grouped by the way they can enter the plant or by where they would find harborages in the plant. If we know the source of an infestation, control is easier.

Insects that live and breed outside the plant but enter occasionally to cause problems include ants, crickets, and earwigs. Insects such as flies are attracted to the plant and may breed inside. Numerous beetles, such as the red flour beetle, are brought into the plant along with the grains or malt. The psocids, such as silverfish, may enter the plant with packaging materials, whereas cockroaches enter through, or live in, floor drains and sewers. Termites could enter through cracks in the floor and may exist almost anywhere in the plant unless control is exercised constantly. Rodents largely come in from the outside and birds roost, rest, or feed near the brewery.

Integration of Sanitation and Pest Control Methods

Although most companies will be dealing with existing facilities, it is important to discuss some factors that would need to be considered when selecting the site for a new plant.

Location and Environment

A number of factors should be considered in addition to the practical considerations of rail and highway access, quality and quantity of water, and labor potential while deciding on the location of a new facility. For example, locating the facility near a populated area may bring complaints of odors, extra traffic and so forth. It will also be increasingly difficult to fumigate any part of the facility, as concerns about fumigant gases increase. A heavily populated low-income area may harbor a variety of pests that will then be attracted to the plant.

There is no practical location that would be pest-free. It is important, therefore, to choose a location where the pest problems are the easiest to identify and manage. If we know ahead of time that flying insects will be the main problem, we can minimize the number of doors and other plant openings, and provide all appropriate physical barriers.

Landscaping to Minimize Attracting Insects

Landscaping must enhance the image of the plant and its product, but minor changes in the landscaping can make a big difference in controlling the pest problems. It is better to avoid fruit and nut trees. The blossoms and fruit will attract honeybees, which may become a problem for some workers. The bees will later head for liquid sugar lines or can-crushing areas. Rotting fruit that has fallen to the ground attracts flies and birds. Nuts can attract rodents.

Structural Design to Exclude Pests

Mechanical design to exclude pests starts with recognition of why pests are attracted inside a brewery. The primary points are:

- Physical shelter from unfavorable weather conditions. Pests may seek the warm building during the winter months or cooler areas during summer months. Heavy rains will also cause pests to enter a building that would not be their normal breeding site.
- Odors of foods will attract hungry pests.
- Presence of a moisture source—most pests need water and this is often lacking outside.
- Lights, particularly white lights over a doorway, attract some pests. Sodium vapor lights would be less attractive but mounting lights at least 10 m (30 ft) from a doorway and shining the light back to areas that need illumination is better.

Perimeter Design

Wall fans with their 'self-closing louvers' often give a false sense of confidence. The louvers do not close tightly enough to prevent insect entrance and, despite many people's insistence, these fans get turned off occasionally. A number of pest infestations have been traced to these fans permitting insect entry during weekends or holidays, when they were turned off.

Doorways are needed but doors must not be left open. All doorways should be designed so that they can be easily closed, and will exclude pests when they are closed.

Metal roll-down doors, when properly installed, are among the best for regularly used doors. However, even these doors will need additional barrier brushes or other material along some edges. The standard steel door and casing is excellent for pedestrian use, but it often comprises of hollow steel sections that can permit mice to run up inside the doorframe.

Air blast fans installed over doorways may help in some situations. They must be properly installed to blow to the outside and must be checked at least semi-annually ensure that they are still operating correctly.

Roof openings are the most neglected insect entrance. Unscreened vents and stacks are left next to leaking dust collectors and puddles of rainwater. Do not underestimate the ability of stored product insects to fly from a nearby grain-processing source to the roof vents of a brewery. Roofs near dust collectors should always be smooth rather than gravel to facilitate cleaning.

Trash and Waste Disposal

Trash compactors seem to be designed for two things—as a convenient place to dump trash, and as a pest feeding area. It is almost impossible to clean thoroughly under the compactors as they are usually placed close to the ground. They are rarely restricted and there is often no drain nearby. If possible, the compactor should be elevated on concrete skids to increase the height under the unit. Of course, there must be a comparable raised area for the truck to pick up the disposal unit. The area surrounding the compactor must be smooth concrete and not asphalt or dirt. There must be a slope to a sewer drain with adequate capacity.

Dumpsters are usually placed near a doorway for convenience. This means that the most pest-attractive area is near a door that is often open or partially open. Electric fly-catching devices should be installed in both areas to intercept flying insects.

Open dumpsters and trash bins placed close to buildings allow pest attack. Moreover, all trashcans should have self-closing lids, or at least tight-fitting lids, and should be lined with plastic bags to facilitate daily disposal and cleaning. A regular cleaning schedule must be followed to assure that all cups and other attractive materials receive proper disposal. Outdoor eating and smoking areas are common at some plants. These areas should be located as far as possible from doorways that lead to production or other critical areas.

Windows that can be opened are unnecessary in a modern brewery. The small amount of light and air movement provided is not worth the potential for insect entry. If light is desired, glass blocks can be used in some areas, but glass in any form presents a risk of breakage and subsequent contamination.

How to Locate Possible Points of Contamination

The modern concept of quality assurance involves a hazard analysis of the potential for contamination at each critical point. Critical points are where contamination may occur if steps are not taken. A trained inspector should do a complete HACCP survey.

Equipment Design

Grain and Malt Area

When the malt or grain arrives, the quality should be rechecked for insects or other contamination. However, even clean rice, corn grits, or malt can be contaminated during unloading.

Unloading at large modern breweries is usually done with a pneumatic system and transfer hoses. When the transfer hoses are left on the ground and not capped, insects and rodents have been known to crawl inside. They will be transferred with the grain. This has happened several times in the past and large amounts of products had to be destroyed. Gravity unloading of railcars into a floor dump cannot be as sanitary as desired. Dirt and insects near the dump may be included with the product. The chance of contamination from stored product insects continues at least through the malt mill and weighing points.

Silos can develop insect infestations that can be transferred from old stock to new stock if a regular cleaning and fumigation program is not followed. Grain and adjunct cleaning systems utilize equipment that is hard to inspect and clean thoroughly. Spot fumigation is often used for inaccessible areas but a redesign of equipment so that all areas can be inspected and cleaned will eventually be needed.

Brewhouse

At the mash and wort areas, the contamination problem largely concerns cockroaches and flies. Wood should be avoided in production areas as it cannot be totally cleaned and is a preferred resting point for cockroaches. Smooth, well-maintained ceramic walls and floors reduce the chance of problems in this area.

Cockroaches often use the sewer and drain system as their own private highway. They hide deep in these areas during 'fogging' operations, with only a few venturing up to be killed. The rest happily feed when the insecticide has dissipated. Cockroaches do not like the light and air movement of a large space.

The floor area around the kettles often contains loose floor tiles or missing grout. This is an ideal breeding area for phorid flies and drain flies as well as cockroaches. An inspection with a pyrethrin aerosol should be done when production permits. Proper sealing is a constant operation.

Filling Area

The filler area is a place where contamination could occur since it is the final point before the product is packaged. The most difficult insect problems involve equipment not currently used. This equipment does not get the same degree of cleaning and inspection, but should receive priority inspection at least monthly.

The filling area presents many contamination possibilities. Part boxes of crown caps are often poorly sealed and can be contaminated with dust or insects. Very few brewers have any cleaning system for crowns. The box is dumped into a funnel-shaped hopper and crowns, bugs, and dusts are funneled to the open containers. It is possible to put an air wash at the feed portion of the crown line. The first line of defense must be keeping boxes of crowns sealed.

The filling line is usually covered. At best, the covers merely protect from something falling in from above. They cannot protect from fruit flies or other insects flying in, particularly during line stoppages. Flying insects must be stopped before they get to this area.

There should be at least two closed doorways between packaging and the outside. Monitoring with traps for flying insects is important in this area. Any appreciable number of insects caught in this area shows a failure at another point that must be addressed.

Insect Control Methods

There is no brewery in the world that does not occasionally have an insect problem. There are a variety of insect control methods and a good pest control program will choose the options that are safest to the product, the employees, and the environment, and integrate these into one master plan. Among the options are:

- Use of a fumigant gas:
  - Fumigate ingredients before they arrive at the plant
  - Fumigate ingredients in bins or silos
  - Fumigate equipment
  - Fumigate packaging materials
  - Fumigate an entire plant or sections of the plant
- Space spraying of large areas with non-residual insecticides
- Limited area treatment with non-residual aerosols
- Baits for ant or cockroach control

- Void treatments with insecticidal dusts
- Crack and crevice treatment with residual insecticides
- Spot treatment with residual insecticides
- Outside bait treatments for flies and other insects
- Nonchemical treatments such as extreme heat or cold

Use of Fumigant Gases

Over the years, fumigant gases have been used frequently in breweries to control insects.

The use of fumigants starts long before ingredients arrive at the brewery. Rice or corn grits and even barley are often fumigated before delivery. It will become even more important to assure that no pests enter with the ingredients as we lose some of the control methods now used in the brewery. Phosphine products are presently the fumigant of choice for grain.

In the brewery, silos are often fumigated when they are emptied to assure that any insects feeding on the dust cling to the sides of the silos or at the bottom of the silos will be killed. This must be done very carefully to assure the safety of unprotected workers who must work nearby during the several days of fumigation. Phosphine products are normally used in this area and they require three of more days to be effective.

Equipment that is difficult to clean thoroughly can become infested in between the fumigations of the entire plant. It is possible to fumigate just the equipment with magnesium phosphide, but this must be handled by specially trained crews.

Fumigation of the entire brewhouse has been utilized by some breweries to assure the lowest possible level of infestation. This is often done several times a year. In some breweries, the cellars are adjacent to the brewhouse and will possibly be contaminated with the fumigant gas. It is very difficult to remove the gas from a cold area and a health hazard could result. This type of fumigation is usually done with methyl bromide so that only twenty-four hours of exposure plus twelve to twenty-four hours of aeration time is required. It is difficult to shut down a brewery for longer periods of time. With the projected loss of methyl bromide other materials will be needed.

Types of Fumigants Used

Methyl bromide has been used extensively for many years because of its many advantages:

- Little or no residue problem
- Good kill of all life stages of insects including the egg stage
- Short exposure time of twenty-four hours plus twelve to twenty-four hours for aeration
- Does not harm electrical equipment
- No significant insect resistance known

Unfortunately, methyl bromide is alleged to destroy the ozone layer and is being banned by international agreement.

No other existing fumigant has all of the attributes of methyl bromide and no new fumigant has been developed with all of the advantages of methyl bromide. We must look at all possible alternatives including other fumigants.

Sulfuryl fluoride is currently being tested as a possible replacement for methyl bromide in food processing plants. It can be used with a twenty-four-hour exposure period and handled in a similar manner to the way that methyl bromide has been used. Tests do not show a residue problem and there is no danger of corrosion unless there is an open flame or equivalent. Although the egg kill is not as good as with methyl bromide, the kill of other life stages is comparable.

Carbon dioxide alone or combined with heat can control insects, but requires longer exposure than methyl bromide. It could have value in some bin treatments, but probably could not be used to fumigate a brewhouse simply because the building usually cannot be sealed adequately to hold the sixty percent or higher concentration of carbon dioxide that would be required.

With any fumigation in a brewhouse or other buildings, safety to employees must be paramount. Even silo fumigations must be done in ways that will protect workers that might travel near the silos.

Space Treatments (Fogging)

Space treatments utilize liquid insecticides dispersed as minute aerosol particles often as small as 5–25 μm in size. When particles are this size, they will float throughout a room for as much as several hours and can contact and kill many different kinds of exposed insects. Space treatments are particularly effective against small flying insects such as fruit flies and other flies, but can kill even large cockroaches when they are exposed.

Space spraying is not the same as fumigation. Fumigant gases move as single molecules and can penetrate cartons, boxes, and even concrete block walls. Fogs can only move between cartons and boxes. The most effective insecticide for space treatments is dichlorvos (DDVP). It formerly was mixed with methyl chloroform (1, 1, 1-trichlorethylene) that helped it disperse as a very fine vapor. This diluent is no longer available and some of the oil-based diluents have had odor and other problems. There are formulations that use carbon dioxide as an aerosol propellant and have had good success.

Baits

Baits for insect control are a very old concept, but recent developments have made them the product of choice for most ant and cockroach control situations. Baits are available in liquid, granular, and gel formulations. When properly used, they do not present the hazards to personnel associated with fumigants or the exposure to contamination problems created by fogging. Although they are initially labor intensive, they are very efficient over a year's time.

For cockroach control, small amounts of a gel or other formulation are injected into the cracks and crevices that are known hiding places for these insects. Control will last for several months or until all of the bait is gone. Baits have always been the first choice for some varieties of ants and baits are now available for most of the types of ants that could be a problem in a brewery. It is usually advisable to use several types of baits, as different ants will prefer different foods. Some ants will even feed on sweet baits occasionally, but prefer protein baits at other times. Fortunately, many types of bait are now available.

Baits are also valuable outside. Granular baits for fly control have been used for many years and can be very useful in trouble spots such as around dumpsters, can crushers, and other areas that are attractive to flies. The flies will normally eat the bait and die in the same area, which seems to attract other flies. Baits should not be used near doorways or other potential pest entrances because you may end up with more flies inside than would normally occur.

Granular-type insecticides are also available to eliminate ants and other insects that can migrate from the outside to the inside. Fertilizer spreaders can be used to apply these materials around the perimeter of the plant. Granular-type materials should never be applied to paved surfaces because birds will eat the baits thinking that they are seeds. Serious bird kills have occurred in this way.

Residual Sprays

We do not have good bait for stored grain insects and they are often the most difficult pests to control in a brewery. Good cleaning and sanitation must be done to aid in the control of stored grain insects. If there is a fine coating of organic dust in many places, there can be lots of stored grain insects and the best pesticides will not work efficiently.

If sanitation is good, residual sprays on the floor or wall areas where stored grain insects have been seen will kill these insects and last for a week or more. These sprays will be more effective if they are placed back in the narrow cracks that harbor these insects. Each insect will have its own preferred type of crack such as the point where equipment is fastened to the floor. Sealing these areas after the crack is sprayed will reduce the required labor over a year's time.

Residual Dust

Dust formulations have advantages over spray applications. They will often have a longer residual life because the toxicant is impregnated on a dust surface known to be compatible with the insecticide. Its major, and probably only use in a brewery, is to treat wall voids or other voids. A wet spray injected into a void such as a hollow block wall would hit the rear portion and drip down. Its repellency action would keep the insect away from the spray residue. Dust will float and coat the entire void if properly applied. New dust formulations are even resistant to moisture problems. Small inexpensive hand dusters are adequate for most uses but small electric dusters are now on the market. Dusting should never be done where the dust can drift and cause a contamination problem.

Heat or Cold Treatments

When temperatures are held to 57–65°C for twenty-four hours, lethal temperatures will penetrate to almost all areas of a building. The cost is usually competitive with normal fumigations. The problem is that breweries are rarely designed to withstand extreme heat. Cellars are often adjacent to the brewhouse and would probably be affected.

Cold storage of sensitive ingredients such as malt samples and hops is a viable practice for controlling pests as well as preserving the quality of the products. The temperatures encountered in cellars would not be lethal but would retard development of most insects.

Insect Monitoring Methods

A thorough flashlight inspection is an indispensable part of any monitoring program but the results can be greatly enhanced with some additional tools.

Pheromones are the natural 'perfumes' that insects use to attract the opposite sex. They have been synthesized and combined with sticky traps. Pheromones can help discover an infestation in its early stages when control is easier and can help monitor progress in control efforts and suggest areas where more work is needed. The most effective pheromone traps so far are those for the Indian meal moth, the cigarette beetle, the warehouse beetle, and other Trogoderma species. At least some of these traps should be placed in any grain storage area where the insects might be found.

There are times when pheromone traps are ineffective. Some areas may have dust conditions that will compromise the sticky trap area. The traps for red flour beetle, confused flour beetle, and saw-toothed grain beetle are not as effective as the others. Newer traps may help with these species.

Glue boards can be placed out of sight in many areas to monitor cockroach and other insect presence. This, coupled with good records, will normally show that over eighty percent of the problems are in less than twenty percent of the total area. Priority can then be assigned to the proper areas.

Electric fly grids have been used as control tools for a long time. Their value can be enhanced if the 'catch' is carefully examined as it can reveal a great deal about flying insect infestations. In addition to the usual houseflies, phorid flies or drain flies may be found, indicating a drain-cleaning problem nearby. Cigarette beetles or dermestid beetles, or various other strong flyers, may indicate a migration of these pests in the area and need for tighter sealing of the plant.

Samples can be removed from central vacuum systems or just the portable vacuum cleaners and sifted for the presence of insects. Insects that were missed with other techniques can be detected this way. Retained samples of incoming ingredients should be checked after a month for any egg hatch that would not have been visible at the time of arrival.

Pest sighting reports by key employees can be one of the best techniques for locating infestations if properly recorded. Normal pest reporting exaggerates the problem or does not pinpoint the location of the observed pest. A typical remark is that "the bugs are all over". This kind of observation does not permit efficient treatment. More specific detailing is needed. The recording form should tell the name of the observer, description of pests seen, location, date and time, and any remarks that would help. A follow-up section of the report should show the action taken by the technician with documentation of a follow-up inspection no more than two weeks later.

Safety

Breweries have always been concerned with the health of their employees—however, in today's litigious society and its media-sponsored fear of pesticides, greater care is needed.

Top priority is naturally assigned to production, and conscious effort from the top executives is necessary to ensure proper precautions are taken. For example, no one should stay in an area where a fumigant or a space treatment is used until the area has been aerated and cleared with proper testing instruments. This may mean that a choice will sometimes be required of whether a routine fumigation will be done or whether an equipment repair will be made. Good monitoring and recordkeeping will help decide the necessity of a 'routine' fumigation or 'routine' space treatment. There is probably too much fumigation or space treatments done that are based on a calendar schedule rather than those that are a real necessity.

The results of tests done one time should never be regarded as 'normal' for all conditions at that brewery. It is important to 'characterize' the actual exposures at each facility under a variety of conditions so that written guidelines can be furnished to the workers.

Treatment of silos inside the brewery should only be done when there is no chance of unprotected persons being exposed to a dangerous level of fumigant. This is very difficult when the gas may be present for three or more days and access to the silos may be in a major travel area. The best way is to rope the area off with 'danger tape' and then monitor it regularly with instruments to record the actual level. Obviously, written records must be kept and any readings above the threshold limit value will require appropriate steps. The use of other gases with shorter exposure times should be considered.

Space spraying presents some hazards. Dichlorvos is considered hazardous and the applicator, if in an exposed area, should wear a self-contained breathing apparatus (SCBA), head covering and body suit that will protect from any exposure.

Pyrethrin sprays may cause allergic reactions in very few people, but it is more important to ensure that employees will not be exposed to either space sprays or fumigants. This will require proper employee notification, placards, and, if possible, locks the keys of which are not generally available. 'Clamshell locks' can be obtained and fitted over door handles.

Careful crack and crevice treatments are not likely to cause any injury to nearby workers, but it is possible to find very small amounts of the pesticide on untreated surfaces days or weeks later because of pesticide volatilization and migration. This is more severe when the spray is applied as a floor/wall juncture spray, which cannot be recommended as a routine application. The small amounts that can be measured would not normally be considered a contaminant at this time but the brew master should be aware that it could occur.

The primary source of information on the safe application of any pesticide is the pesticide label. The manufacturer should be contacted on any point that is not clear. Material Safety Data Sheets (MSDS) are also valuable, but since they are written for everything from the formulation of the insecticide to its use, they may be overly restrictive for actual use in a brewery.

All technicians or other potentially exposed persons should have a thorough medical examination prior to their first pesticide exposure so that baselines for each individual can be established. Blood cholinesterase tests are needed before exposure to carbamate or phosphate pesticides. No one should be required to wear a respirator until a doctor has certified through examination that the person is capable of working while wearing that type of respirator. A physician must check any symptom of pesticide poisoning.

Perimeter Rodent Control

Careful examination of all incoming products is required under the Good Manufacturing Practices regulations and will have eliminated most cases of pest entry with products but most rodents enter from the outside. They are excellent climbers. If downspouts are not screened, rodents can use them to get to the roof areas. They can also climb rough walls or jump from nearby trees or other buildings.

Excluding rodents is the primary control method. Doorways should be checked at least monthly for any crack that would permit rodent entrance. As a rule-of-thumb, if a pencil can slide under a door, a small mouse can enter. If your thumb can slide under the door, even a rat can enter. Mice can squeeze through holes as small as 4 mm.

The gravel strip around the building or paved areas next to the building are a very important part of the rodent control measures. Rodents are shortsighted and cannot see a doorway or other opening from a distance. They must creep along a wall until they find an entrance. As they are constantly afraid when they do not have the protection of tall weeds, they will enter any shelter, including rodent bait stations. If there is fresh clean bait, they will usually sample it and die before they can enter the plant. The best plan for rodent control will have two or more lines of defense. In addition to the bait stations next to the wall of the plant, there should be another row along the perimeter fence line.

Experience and good records will determine the appropriate number of bait stations along perimeter fence lines. If there is a grain processing plant, neglected buildings, or similar suspected harborage nearby, more bait stations should be installed to intercept this possible source. Bait stations with granular baits or wax blocks are effective for house mice, but are only marginally effective for roof rats, which prefer fruit or meat baits rather than cereal baits.

Roof rats can enter the plant without crawling along the walls. They can jump from a tree limb to a roof when the distance is as much as a meter. It is therefore important to plant trees away from the building or at least keep them pruned to over a meter from a roofline. Roof rats can also crawl along power or telephone wires and enter the plant through the small openings where the wires enter. Ideally, the power lines should be buried so that this will not be a problem. If the rats are using the wires for an entrance, they can still be stopped. A heavy sheet of plastic formed into a tube around each wire can be sealed with tape and will hold for years. If the rodent steps on the tube, it will turn and the rodent will either back off or fall off.

All bait stations should be 'tamper-resistant', and should have checks to prevent a child from easily removing the bait. Breweries are very sensitive to adverse public opinion. The information that children obtained rat poison from your facility is not good for public relations. Such tamper-resistant bait stations are expensive and not foolproof, but they are mandated by the EPA on most rodenticide labels. Even these bait stations must be securely closed with a lock or some other method that will make it difficult for a child with a pocket knife to open the bait station. Special screws are used on the lids of some commercial bait stations. As only a special tool will open them, there is a reduced chance of pilfering.

The bait stations must also be secured to the ground if there is any possibility that a child or non-target animal could have access to it. 'Liquid nail' or similar glue can be used to glue a bait box to a concrete area. On grassy or gravel areas, the bait box can be fastened to a large terrace block. Long spikes may be sufficient on asphalt-paved areas. There are also commercial anchoring devices.

All bait stations should be inspected and cleaned at least twice a month and the bait should be replaced at least once a month. Recording stickers are often placed inside the bait station to record each cleaning operation to assure consistent service. Clean stations with fresh bait are far more effective than dirty stations with moldy insect-infested bait.

Multiple catch traps can be used outside as well as in strategic areas inside, but will need protection from damage due to traffic. Traps or stations with glue boards are not subject to the fastening restrictions, but this still may be desirable to prevent stealing. There will rarely be cases where baits will be needed that are stronger than anticoagulants or similar materials. Zinc phosphide-coated baits may be needed as one-time treatments in special situations.

With a well-maintained program, toxic baits will not be needed inside the building. However, multiple catch traps, covered glue boards, and probably some snap traps will be needed to catch the stray rodent that gets inside the building. These traps should be checked at least once per week.

The perimeter should be inspected at least once a month for any sign of rodent burrowing. This will often occur near drainage ditches or under concrete pads surrounded by dirt. Additional bait stations may be needed when these burrows are found or it may be possible to use phostoxin tablets or wax blocks directly in the burrows if the labels permit and the technician is certified. Use of phostoxin tablets requires a fumigation certification and training. The numerous reasons why rodent control measures might fail are summarized in Table 14.1.

Table 14.1: Some Reasons Why Rodent Control Programs May Fail

Plant environment
- Poor sanitation results in excessive rodent populations
- Weeds provide shelter and food
- Trash and stored equipment provide shelter
- Spills along railroad tracks provide food
- Poor drainage provides water

Outside bait stations
- Not in the travel path of rodents that are entering plant
- No bait or insufficient bait for rodent population
- Old dirty bait that is no longer attractive
- Bait station poorly designed and is not an attractive shelter
- Stored materials and trash provide better shelter
- Bait station so hot or so cold that rodent does not enter
- Rodent resistant to bait used

Rodent proofing
- Some openings not closed
- Openings not sealed with correct materials
- Doors left open during the day
- Rodent proofing at ground level only
- Rodent proofing at doors but not around pipes and other entrances

Inspection of incoming ingredients
- Little or no inspection at the time of arrival
- Open truck not inspected until unloading starts (rodent may have left)
- Shrink wrapped material assumed to be clean—receives no inspection
- 'Chimney-packed' pallets not checked in center
- No backlight used even on products preferred by rodents

Multiple catch traps
- Animal (rat) too large for trap and no other traps used
- Trap damaged and not inspected. It cannot catch mice
- Trap wound too tight to catch small mice
- Trap too far from wall to be effective
- Trap left too long in one spot and rodents avoid it
- Trap not cleaned and has bad odor

Baited snap traps
- Poor choice of bait; undesirable food for rodents
- Bait not tied on; easily pulled off without triggering trap
- Trigger of trap not against wall
- Trap warped and wobbles when rodent touches, scaring rodent
- Trap sprung by vibration of plant before rodent gets there
- Trap not left out long enough to overcome fear of new objects
- Trap not tied down and is dragged off to where rodent can get free
- Trap stored with insecticides and is repellent to rodent
- Prebaiting not done on 'smart old rat'
- Trigger angle too high or too low

Glue boards
- Poor placement in relation to rodent pathways from shelter to food
- Glue layer too thin for size of rodent
- Glue has a layer of dirt, making it ineffective
- Placed in a moist area where rodents wet feet may make board fail
- Board is not fastened and is dragged to where rodent can pull it off

Inspection program
- Plant inspections infrequent
- Inspections only made of inside areas
- Inspections made by untrained person
- Inspector unwilling to get dirty during inspections
- Inspector thinks in terms of chemical control only

Regulations Affecting Pest Control

Laws and regulations concerning the application of pesticides are written at the government level in India. It is usually under a department of agriculture, health, environmental protection agency, or an independent body. Other agencies, such as the Occupational Safety and Health Administration (OSHA) and the Food and Drug Administration (FDA), have regulations that must be followed.

Regulations will differ from state to state but, basically, pesticides can only be applied in a manner consistent with its label directions and only materials registered with EPA may be used as pesticides.

Fumigants and other "restricted use pesticides" can only be applied by or under the direct supervision of a "certified pesticide applicator". The certified applicator should have passed a state exam to show competence. This certification will need to be renewed on a regular basis through examination or by attendance at special training meetings. Fumigants can be applied and aerated only by a certified individual. Work in a brewery may require general pest control, termite control, fumigation, bird control, and possibly lawn and ornamental pest control or weed control certifications.

When an outside contractor is used, the brewery should have copies of the certification cards of all personnel who will be working in the plant. Even when an outside contractor does all of the pest control, it is suggested that the brew master or other responsible person also hold a certification. This will help in administering the program and understanding the implication of all the laws.

Recordkeeping

Breweries should maintain files or records of all pesticide applications involving the use of restricted use pesticides is kept for at least ten years after the application of all pesticides.

Pesticide records should be kept by either the outsourced pest control agency or company and by the brewery. Pesticide records should not be on the same form that reports pest sightings or sanitation problems. There is no reason to open up this kind of information for examination by other people.

Pesticide use reports should contain at least the following information:

- Name of pesticide and registration number
- Amount used
- Location
- Target pest
- Method of application
- Date applied

In addition to the records required by law, there are records that can facilitate a good pest control program. All pest control programs in a brewery should have a written list of what is to be done and when. Good recordkeeping will track the progress of the program and help avoid any problems. Any change should be done only with written advance permission from the brew master or other responsible person. Rodent catch records can help evaluate success of the control program and point to any modifications needed.

Computerized tracing and recording programs are now available. Maintenance requests that involve sanitation problems should be logged separately and checked to ensure that they are performed in a timely manner. Any remodeling or construction done in or near the plant can cause new pest problems or disclose old, obscure ones. Each construction program should result in notification of the person in charge of pest control so that an evaluation can be made of extra steps that may be needed. Poor remodeling practices can create many new entrances for pests. Entrance can be gained from the outside through wall voids left by careless repairs where pipes and electrical wiring were brought into a building.

Outside Contractors versus In-house Pest Control

The brewing industry primarily uses outside contracting firms of pest control who have experience in other food plants. Exceptions are usually made for a person with a college degree in entomology or a related science. As it becomes more difficult to replace certified persons, there will probably be greater dependence on outside contractors.

Some pest control companies are recognizing this trend and are training people across the country to specialize in work in the more demanding field of pest control in a food plant. A firm that merely treats residential property would not have the expertise or the knowledge of appropriate laws to handle a brewery account.

A qualified outside contractor can offer many advantages:

- Certified staff
- Access to all types of expensive application equipment
- Employees that regularly attend training meetings and know the latest techniques and pesticides
- All pesticides stored off your premise

- Separate insurance and responsibilities in case of an incident

- Not involved in any company politics or rivalry

- Personnel experienced in specialized aspects of pest management such as commodity fumigation; (not all pest control companies are qualified to work in a brewery or any other food plant). Poorly qualified companies can cause the following problems:

  o Turnover of technicians can mean workers not familiar with your plant or your safety procedures

  o Employees on a commission may try to rush through the large accounts, such as a brewery

  o They may not be properly trained in pest control much less pest control in a brewery

  o The technician answers to his boss and not directly to the brew master

  o The technician may try to hide the extent of infestations so that he will 'look good'

  o Some companies may have excellent technical staff, but unless they visit you on a regular basis, they are of little value

A successful pest control program starts with detailed specifications, usually written by the corporate brewing staff but often with the help of a specialized consultant. The potential for alleged product contamination is always possible, regardless of who does the work. Contamination problems or other adverse publicity will fall more heavily on the brewery than on the pest control company. The brewery does not step down from its responsibilities by hiring an outside contractor. There must be adequate recordkeeping assuring the brewery supervisor that the work has been done as planned. There should be quarterly reviews of the progress of the program between supervisory personnel of the pest control company and the brewery staff. Poor communication is often the reason for problems between the two.

Evaluating Pest Control Results

There are various ways to evaluate the results of a pest control program. Obviously, if complaints have dropped to near zero this may be one criterion, but it could give a false sense of security. There should be regular in-house and contract inspections to monitor the progress. The in-house inspection should make a weekly check of key areas and a month-wise recorded inspection of the entire plant. Records of pheromone and glue board trap catches should be integrated into the report.

The contract inspection can be made by the corporate staff or one of several excellent outside agencies. It is important when an outside agency is hired to tell them ahead of time what is expected and any company safety rules that may affect the inspection procedure. The highest corporate officer should see all major reports and at least a summary of the weekly reports.

There are several reasons why inspection programs fail:

- Plant inspections are infrequent
- Inspections made only of inside areas
- Inspections made by untrained personnel
- Inspector unwilling to get dirty during inspections
- Inspector thinks in terms of chemical control only

Summary

Insect and rodent control has changed in several ways and will change in even more ways in the future. It is important to recognize the principles behind each pest control technique.

The first group is made up of insects that live primarily outside and enter only occasionally. Control of these will always depend on good sanitation and avoiding any vegetation close to the building. This keeps their natural feeding and resting area further away from potential plant entrances. At the present time, a number of insecticides are available to provide residual control of these insects as they try to find entry points. Most of these will need to be applied on a monthly basis during the warm months if a problem exists. It is always better to inspect and monitor with glue boards to determine if control is needed.

The second group comprises flying insects such as flies. Again, sanitation is key to their control, but it would be hard to keep the brewery and grounds so clean that flies would not be attracted. Careful use of space sprays during times when that part of the brewery is not in operation will often be needed. Synergized pyrethrins are currently the first choice. If small flies are kept under control, there will not be a problem with spiders and their webs.

The third group is made up of those insects that enter with packaging. In the future, more attention will be paid to non-pesticidal controls such as plastic slip-sheets or low-starch slip-sheets. Fumigation will always be an option but there will be a trend away from fumigation whenever possible.

Roaches, ants, and even termites will be controlled with baits. Some of these baits are in use now but even more are being developed. Baits are the least likely to cause contamination and are increasingly cost-effective.

Rodent control will probably change less than insect control but will still rely on sealing the buildings so that rodents cannot enter.

15

Brewery By-Products and Effluents

Introduction

Beer is about ninety-five percent water in composition; however, the amount of water used to produce a container of beer is far greater than the amount of water contained in the beer that is actually packaged and shipped out. Water usage varies widely among breweries and is dependent upon specific processes and locations.

In addition to the water used in production, wastewater generation and disposal presents another improvement opportunity for brewers. Most breweries discharge seventy percent of their incoming water as effluent—which is defined as wastewater that is generated, and flows to the sewer system. In most cases, brewery effluent disposal costs are much higher than water supply costs.

Under present conditions, and with the way water is being managed, we will run out of water long before we run out of fuel.

Typical Brewery Water Use per Area

Department	Water Usage
Brewhouse	25%
Cellars	17%
Packaging	38%
Utilities	20%

Within a brewery, there are four main areas where water is used:

1. Brewhouse
2. Cellars
3. Packaging
4. Utilities

In addition, ancillary operations such as canteen and restrooms contribute to water usage.

Main Areas of Wastewater Generation

Source	Operation	Characteristics
Brewhouse		
Mash Tun	Rinsing	Cellulose, sugars, amino acids. ~3,000 ppm BOD
Lauter Tun	Rinsing	Cellulose, sugars, spent grain. SS ~3,000 ppm, BOD ~10,000 ppm
Spent Grain	Last running and washing	Cellulose, nitrogenous material. Very high in SS (~30,000 ppm). Up to 100,000 ppm BOD
Wort Kettle	Dewatering	Nitrogenous residue. BOD ~2,000 ppm
Whirlpool	Rinsing spent hops and hot trub	Proteins, sludge and wort. High in SS (~35,000 ppm). BOD ~85,000 ppm
Cellars		
Fermenters	Rinsing	Yeast SS ~6,000 ppm, BOD up to 100,000 ppm
Storage tanks	Rinsing	Beer, yeast, protein. High SS (~4,000 ppm). BOD ~80,000 ppm
Filtration	Cleaning, start-up, end of filtration, leaks during filtration	Excessive SS (up to 60,000 ppm). Beer, yeast, proteins. BOD up to 135,000 ppm
Beer spills	Waste, flushing, etc.	1,000 ppm BOD
Packaging		
Bottle washer	Discharges from bottle washer operation	High pH due to chemical used. Also high SS and BOD, especially through load of paper pulp
Keg washer	Discharges from keg washing operations	Low in SS (~400 ppm). Higher BOD
Utilities		
Miscellaneous	Discharged cleaning and sanitation materials. Floor washing, flushing water, boiler blow down, etc.	Relatively low on SS and BOD. Problem is pH due to chemicals being used

In addition, ancillary operations such as offices, canteen and restrooms contribute to water usage.

Water awareness and conservation practices provide an effective mechanism for brewers to reach out to communities. Outreach efforts have a number of benefits, including building brand image and being recognized as an important part of the community.

In the process of brewing and packaging of beer, the generation of by-products and waste products is unavoidable. Technological advances and improved microbiological control have enabled the brewer to reduce product losses and to produce valuable by-products from materials that were previously considered waste products. Economic advantage derived from minimizing product losses or upgrading waste products to by-products.

Brewhouse Effluents

Effluents from the brewhouse discharged to the sewer include rinses from the various brewhouse vessels, CIP solutions, brew kettle vapor condensate, and liquor from wort clarification that is too turbid to include with the wort.

After run-off of wort to the kettle is completed, the brewer's grain is allowed to drain while the free liquor is collected or sewered. Alternatively, the brewer's grain is immediately conveyed 'as is' to the brewer's grain processing area. The exact procedure depends on brewing practice and on the wort clarification device used.

In a Lauter Tun, which is the most commonly used clarification device, free liquor is usually first drained to a separate holding vessel until this so-called sweet water becomes too turbid. At this point, it is diverted to the sewer. After it has drained to the sewer, the wet brewer's grain still contains about seventy-seven to eighty-one percent moisture. After discharging the wet brewer's drain to a holding tank, the area under the false bottom is rinsed. The rinse water, which may contain a considerable amount of suspended solids, is flushed to the sewer.

Collection and recycling of sweet water is done to improve lautering efficiency. If the lautering efficiency is already high, such as in low-gravity brewing or when adequate time is available for lautering, the dissolved solids concentration of sweet water might be too low (less than 0.8°P, for example) to be economically attractive. Other factors that may make recycling of sweet water unattractive are higher than normal concentrations of suspended solids and soluble b-glucans, which could impede run-off. Proper microbiological control of sweet water must be done if recycling is employed. If sweet water is not recycled, all free liquor from the Lauter Tun is sewered.

Different methods employed at various breweries cause the volume of recycled sweet water to vary from 0 to 6 hl for each 100 hl of final product. This sweet water may have an extract concentration ranging up to 3°P.

It is estimated that the total volume of Lauter Tun effluent and rinses discharged to the sewer varies between 4 and 12 hl for each 100 hl of final product. The dissolved solids concentration in this effluent may vary from 0.4 to 3°P whereas suspended solids concentration may range up to 1.0 weight percentage. For each 100 hl of final product, a typical brewery will discharge about 11.6 kg of dissolved solids and 3.9 kg of suspended solids to the sewer. The BOD of this effluent is approximately 10.1 kg.

Other brewhouse effluents are rinses and CIP solutions from brewhouse pipes and vessels. Heating surfaces in the kettle or in external boilers become quickly fouled by a build-up of proteinaceous material. These surfaces may require cleaning after every two brews. Similarly, wort coolers may be CIP-ed every three to four brews. Wort may also end up in brewhouse effluent through:

    a. Entrainment by brew kettle vapors

    b. Purposely sewering wort to avoid brand mixing

    c. As part of hot wort trub that is not recycled or used as by-product

    d. By leakage and spillage

Brewery Solid Wastes

The most cost-effective method for significantly reducing effluent load of brewery wastewater is to separate the solid wastes from the wastewater itself. The equipment necessary includes holding vessels, tanker trucks that can haul away the material, pumps, and dedicated piping or hoses for transfer. Typical solid wastes include spent grains, trub, spent yeast, diatomaceous earth slurry from filtration, and packing materials.

Spent Grains

Beer production results in a variety of residues, such as spent grains, which have a commercial value and can be sold as by-products for livestock feed. The nutritional value of spent grain is much less than that of the same amount of dried barley, but the moisture makes it easily digestible by livestock.

Spent grains are used as a constituent of cattle food and the food value of 5 kg of moist spent grains is equivalent to 1 kg of barley. However, they are bulky and begin to decompose quickly, so they must be removed from the brewery promptly. The composition reported, on a dry weight basis are:

- Crude protein   : 27%
- Fat   : 6–7%
- Ash   : 4–5%
- Crude fiber   : 15%
- N-free extract   : 46%

The moisture content of spent grains varies widely depending on the wort separation system used. Thus grains from a Strainmaster may contain eighty-seven to ninety percent moisture, and they are sloppy and easy to pump, but they are so wet that liquid drains from them and they must be dewatered before removal. Water drains from grains with moisture content above eighty percent. The drain water is an excellent medium for unwanted microbes. Grains from Lauter Tuns can contain seventy-five to eight-five percent moisture and those from pressure filters contain as little as fifty to fifty-five percent.

A range of other uses for spent grains has been proposed, such as:

1. As a source of biogas and soil conditioner produced by anaerobic digestion
2. Disposal by burning (giving heat)
3. As a source of 'secondary worts' generated by acid or enzymic hydrolysis
4. As a source of protein
5. As a source of food grade fiber
6. As a basis for mushroom compost
7. As a soil conditioner and organic fertilizer
8. As a medium for growing earthworms to use in poultry food, and in fish food

Trub

Trub is slurry consisting of wort, hop particles and unstable colloidal proteins coagulated during the wort boiling.

The main component of the trub is coagulated proteinaceous material formed in the brewing kettle—some also develops during mashing. The amount of trub formed depends on many factors—protein content of the malt, the amount of protolysis during malting, kilning conditions, mashing schedule, polyphenol contents of malts and hops, method of boiling (internal vs. external calandria), length of boil, oxidation during kettle boil, and hopping method.

Trub is separated from wort by sedimentation in a conical bottom hot wort tank or in a whirlpool tank. Even in a whirlpool tank, the sediment still contains considerable wort and recovery of this wort can increase brewhouse yields from 0.6 percent to 1.5 percent.

Recovery of wort from trub is accomplished using a filter press, a vibrating screen, a centrifuge, or by recycling the trub to the top of the grains in a Lauter Tun prior to sparging. The latter method is simple but has a disadvantage—it can only been employed when the Lauter Tun is processing the same type of wort at the time of recycle. Recycling trub also has another disadvantage—it slows run-off and decreases the efficiency of wort extraction. Recovery of wort by means of a decanting centrifuge has been successful and has none of these disadvantages. A new method that is currently being developed is recovery of wort from trub by means of cross-flow filtration.

When wort recovered from trub is fed forward, it is important to control the level of residual suspended solids. An increase in wort suspended solids causes a more vigorous fermentation; the solids either serve as nucleation sites for CO_2 bubbles or yeast growth is stimulated by unsaturated lipids and zinc in trub solids.

Trub is generally mixed with brewer's grain; both have similar amino acid content. Trub contains thirty percent digestible crude protein, which is about twice the level found in brewer's grain.

The addition of trub to brewer's grain will therefore enhance its nutritional value. Trub may also be sold in mixtures with yeast and centrifuge solids recovered from brewer's grain liquor. This type of mixture can have protein content close to that of soybean meal and may be excellent liquid feed for pigs.

Spent Yeast

In brewing, surplus yeast is recovered by natural sedimentation at the end of the fermentation and conditioning. Only part of the yeast can be reused as new production yeast. Spent yeast is very high in protein and B-vitamins, and may be given to livestock as a feeding supplement.

In typical lager fermentation, about 0.27 kg of surplus yeast solids is produced per hectoliter of product. These solids include pure yeast solids, beer solids and trub solids. Brewer's condensed solubles are produced by concentrating press liquor and other carbohydrate-rich brewery waste streams; multiple effect evaporators are generally used for concentration. The product is blended with brewer's grain or is sold as a high-energy liquid feed ingredient, as a pellet binder, or as a feedstock for fermentation.

Diatomaceous Earth (DE) Slurry

Diatomaceous earth slurry from the filtration of beer also constitutes a very large category high in SS and BOD/COD. Different methods for regeneration are under development, but presently they are not capable of totally replacing new diatomaceous earth.

Packaging Materials

Other solid wastes include label pulp from the washing of returnable bottles, broken glass (cullet), cardboard, bottle caps, and wood that is usually disposed of at sanitary landfills.

General Characteristics of Brewery Wastewater

Breweries produce a large quantity of highly polluting wastewater that is rich in organic substances. As the scale and production of the brewing industries increases, the amount of wastewater increases substantially, resulting increasing pollution problems in the environment. Due to wide varieties in its strength (in terms of COD, BOD, and TSS), pH, and the amount of wastewater discharged, brewery wastewater tends to be very difficult to treat. In view of this situation, there is a need to develop technology that is capable of efficiently treating the increasing volumes and strength of wastewater from the brewery.

The wastewater discharged from the breweries is generally a combined effluent comprising discharge from various sources within the brewery. The fermented liquor is the final product in the brewing process. Waste arises from the separation of grain residues, from spent hops, hot and cold break and yeast from the fermented processes, from spillage, from possible spoilage of beer, from the fillers as well as from packaging and from washing wastewater. The wastewater production rates from the brewing and packaging departments vary independently of one another. While the packaging process produces a high flow, high pH, weak waste composed primarily of spilled beer and caustic cleaning solutions, the brewing produces a low flow, neutral pH and high strength alcohol-carbohydrate-protein waste.

A continuous monthly monitoring of the effluent from a brewery showed considerable variation in general wastewater characteristics in terms of biochemical oxygen demand (BOD), chemical oxygen demand (COD), and solids concentration. As described in Table 14.1, total BOD varied from 87 mg/l to 6550 mg/l, and concentration of suspended solids from 16 mg/l to 1162 mg/l. The ratio of soluble BOD to total BOD was about 0.91, which implied that most of the biodegradable materials were in soluble forms. One important characteristic of brewery waste is its fluctuations in flow and quality at night and weekends, compared with average working day flow. The fluctuation in the quantities of wastewater discharged from a brewery is depicted in Figure 14.1.

Table 14.1: The general characteristic of brewery wastewater

Total BOD	41–4260 mg/l
Soluble BOD	34–3890 mg/l
Total COD	87–6550 mg/l
Soluble COD	37–4830 mg/l
Total Suspended Solids	16–1380 mg/l
Volatile Suspended Solids	11–1230 mg/l
pH	6.1–9.1

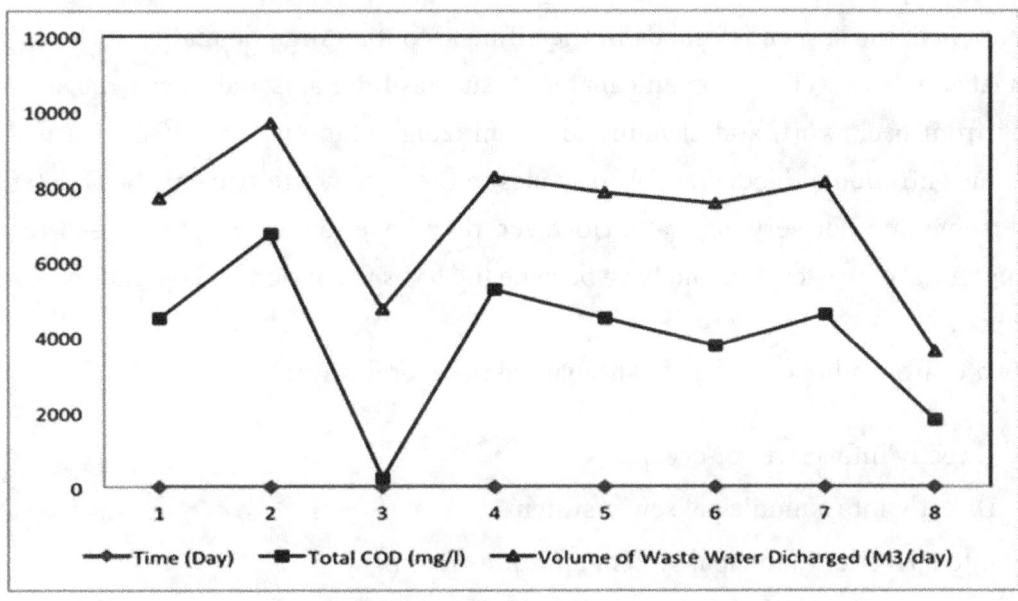

Figure 14.1: Monitoring of brewery wastewater

225

The pH of the wastewater from various processes within the brewery is neutralized in a pH tower within the brewery. This ensures that the pH does not change significantly. Since brewery wastewater has a poor buffering capacity, hydrolysis and anaerobic activity usually reduce the pH. The pH tends to drop from 10 to 4 within a day at room temperature, and it will drop from 10 to 5 within three to four days in a walk-in cooler at temperature approximately 4°C. Due to preponderance of the carbonaceous matter, it tends to be relatively short of nitrogen nutrients. Slight seasonal temperature variations in the wastewater range between 19°C in winter and 39°C in summer.

Brewery Wastewater Treatment

Due to its high organic content and high biodegradability, brewery wastewater is ideally suited to biological treatment. All treatment methods basically involves the conversion by microorganisms of fairly complex stable/unstable organic compounds to CO_2 and water. Biological treatment results in the removal of BOD, the coagulation of non-settleable colloidal solids, and stabilization of organic matter

Wastewater Disposal and Treatment

Introduction

Brewery wastewater is relatively simple and is highly biodegradable. A complicating factor, however, is that wastewater volumes, pH, and concentrations of included solids vary constantly. In order to get a good characterization of the wastewater, flow rates and concentrations must be measured simultaneously over an extended period of time. The flow proportional sampling method is most suited for this purpose.

As mentioned, the suspended solids in the effluent contain organic matter such as grain, trub, yeast, and label pulp as well as inorganic materials such as filter aids and silica gel. Dissolved solids are mainly from beer, wort, and cleaning and sanitizing solutions. The BOD is usually used to index the concentration of biodegradable organics in brewery waste streams. BOD determinations are cumbersome and not very accurate. However, they have historically been used to assess the pollution potential of wastewater and have become the basis for design and operation of wastewater treatment plants.

Wastewater from a brewery may be discharged in several ways:

    a. Directly into a river or ocean

    b. Directly into a municipal sewer system

    c. Into a river or municipal system after pre-treatment

    d. Into the brewery's own wastewater treatment plant

Discharges into public waters are often subject to limitations in organic load, suspended solids, pH, temperature, and chlorine. The maximum allowable BOD limit for discharge of effluents into public waters near densely populated areas can be 20–30 ppm and may be as low as 10 ppm. It is mandatory for breweries to construct a complete wastewater treatment facility to treat their effluent. The high costs that are often required for waste treatment offer breweries an additional incentive to eliminate unnecessary waste and to optimize the reuse of effluents.

Wastewater Volumes and Concentration

As discussed, discharge of BOD and suspended solids from various areas of a brewery may vary considerably, depending on waste management in a large brewery.

Concentrations of BOD and suspended solids vary considerably, depending on where samples are taken and how much water is used for dilution. Many brewers have made a concerted effort to reduce water usage in order to lower costs for water, water treatment chemicals, and wastewater treatment surcharges based on flow. Not all wastewater requires treatment. Non-contact cooling water and rinse water for non-returnable bottles and cans, for example, is relatively clean and may be discharged directly into a river or storm sewer depending on temperature and chlorine limitations. The volume of wastewater that is discharged into a sanitary sewer or treatment plant from large modern breweries is in the range of 2.5–6 hl/hl packaged.

Volumetric discharges are usually higher for smaller breweries, or for older breweries, or breweries located in hot climates.

Wastewater Pre-treatment

Most large breweries require some degree of wastewater treatment, whether their effluent is discharged to public waters, to municipal treatment systems, to their own aerobic or anaerobic systems, or used for land application. In many cases, mere pre-treatment might be sufficient to meet local regulations. Pre-treatment is done by physical, chemical, or biological methods, or by combinations of these.

Physical pre-treatment methods for breweries include coarse suspended solids removal, flotation, and sedimentation. The first step in pre-treatment is usually screening to remove coarse suspended solids such as labels, caps, glass fragments, plastic and other scrap materials, and grain particles.

After screening, the effluents are usually passed through a rectangular grit removal chamber equipped with a continuously operated scraper mechanism. Grit chambers can also serve as a mixing chamber for pH control systems and as a pre-aeration unit to prevent anaerobic conditions in the primary clarifier.

Major problems associated with treating brewery effluents include the highly variable nature of flow, BOD concentration and pH. The most efficient action that can be taken is the installation of an equalization tank, which will be most effective when the contents are mixed well.

Aeration or oxygenation may be required to prevent microbial production of hydrogen sulfide odors. It is advisable to collect the caustic CIP and bottle washer effluents in a separate tank and gradually meter the contents into the treatment plant influent.

Dissolved air flotation is an effective pre-treatment method. In this process, air is dissolved in water under pressure and comes out of solution as tiny bubbles to that particle attach. To obtain large flocs, metal coagulants and polyelectrolytes (flocculation agents) are used. Acids, caustic, and DE are sometimes used to aid in sludge dewatering in a filter press.

Chemical pre-treatment is used in a number of breweries by neutralizing caustic effluents from CIP systems and bottle washers with waste CO_2. Neutralization with sulfuric or hydrochloric acids is usually not recommended because of their corrosive nature and sulfate and chloride discharge limitations. Waste CO_2 sources may be used. Packed trickle flow beds can be efficiently used with good control.

Biological pre-treatment systems include aerobic systems (with short resident time) or anaerobic systems. BOD may be reduced up to sixty or seventy percent—which may be all the treatment that is required, or the first stage in a full treatment process. Aerobic systems used for pre-treatment can be tanks with plastic support media for biological growth such as bio-disks, or it can be as simple as an aerated equalization tank. Some type of primary clarifier is required for sludge removal.

Aerobic Treatment Systems

Most biological treatment systems used in breweries today are aerobic activated sludge systems, although the trend since the mid-1980s has been toward anaerobic treatment systems.

Newer aerobic systems use plastic media that support biological growth. These media can be rotating sheets as in the rotating biological contacter, self-supporting packing materials as in bio towers, or low-density material kept in suspension by fluid motion (fluid bed systems). All of them require less space than conventional activated sludge plants, as they provide more efficient contact between wastewater and sludge.

In the rotating biological contacter, a series of plastic sheets is mounted on a shaft. The closely spaced sheets, 1 cm apart, rotate in wastewater at about 1.5 rpm and are about forty percent submerged. The biological growth on the sheets drags wastewater along, which is aerated as the sheets emerge from the liquid. Excess build-up of biomass sloughs off and is carried out with the liquor to a clarifier.

A bio tower is basically a packed bed with a plastic packing material of high specific surface area and porosity. Biomass grows as a thin film on the packing while wastewater trickles over the biomass. The wastewater is usually recycled to obtain sufficient flow for uniform wetting. Unpleasant odors may be a problem with these systems.

A fluid bed system is probably the most efficient biological growth system as a very high density of biomass can be kept in suspension. Another advantage of this system is that a settler is not required—excess sludge is recovered directly by mechanical removal of the sludge-laden sponges from the tank, squeezing the sludge out, and returning them to the tank.

Another novel aerobic system requiring only a small land area is the deep shaft aerator system. In this process, wastewater is contacted with biomass and compressed air in an underground shaft 152 m deep, which promotes a high rate of biological oxidation.

Anaerobic Treatment Systems

Anaerobic digestion is a complex process in which a large variety of bacteria break down organic materials and convert them into methane and CO_2 in a ratio of about 3:1. The bacteria are generally grouped into three basic types.

The first group consists of acid-forming bacteria called acidogens, which provide extracellular enzymes that hydrolyze soluble and insoluble complex organics and convert the hydrolysis products into fatty acids, alcohols, CO_2, ammonia, and hydrogen.

The second group, called acetogens, transforms the products from the first process into acetic acid, hydrogen, and CO_2. The third group consists of methanogens, which convert acetic acid, H_2, alcohols, and some of the CO_2 into methane.

As a result of recent advances in the basic biochemical understanding of the process, anaerobic systems have found a wider acceptance as biological treatment systems for brewery effluents. Advantages of the anaerobic system over aerobic treatment mentioned in the literature are:

1. Low energy consumption especially in hot climates

2. Net producer of methane gas, which can be used as a fuel

3. Seventy percent less space requirements

4. Less nutrient requirements

5. Up to ninety percent less sludge production

6. Lower capital cost because of a higher loading rate

A number of potential disadvantages for anaerobic systems are:

1. The slow growth rate of methane forming bacteria

2. Anaerobic systems are more sensitive to shock discharges than aerobic systems and generally require a balancing tank

3. BOD removal rates vary and may not always be sufficient

4. Besides seventy to eighty percent methane, biogas is saturated with water and contains twenty to thirty percent CO_2 and 1000–2000 ppm H_2S, which may cause problems in steam boilers

A typical design for an anaerobic digester is a packed column in which the biomass is attached to plastic media. A system of this type with a down-flow mode has been successfully applied to treatment of distillery wastes.

The most commonly employed anaerobic system in breweries is the *Up-flow Anaerobic Sludge Blanket* (UASB) system. In this system, the influent is uniformly distributed over the bottom of the reactor and is then passed through a bed of active anaerobic sludge. Gas, liquid, and sludge are separated from each other above the bed. Most of the sludge returns to the bed, while a portion may be diverted to the surplus sludge tank. Difficulties with this system have been excessive H_2S formation and the loss of solids from the sludge bed as a result of shock loads.

A good solution to these problems has been the design of a two-stage process. The first stage is a stirred tank reactor where acid formation takes place and the second stage is a UASB reactor where methane formation takes place. This two-stage system was found to be quite stable and COD removal rates of seventy percent and higher was obtained at temperatures as low as 15–20°C. Problems with H_2S have been solved by removal of H_2S from biogas by silica gel adsorption or by the addition of iron to form iron sulfide. The newest design of anaerobic digestor, which appears promising, is the anaerobic fluid bed reactor. In this system, the biomass is attached to an inert carrier that is maintained in a fluidized state by an upward flow of wastewater. The main advantage of this system is that a high biomass concentration can be maintained, resulting in a very compact system. A two-stage system has been employed for the treatment of a yeast plant effluent, with a fluidized-bed reactor for each stage. As the biogas from this system contains one to two percent H_2S, it is highly corrosive and reactors are made of glass fiber-reinforced polyester with PVC lining.

In general, anaerobic treatment systems work well for effluents with high BOD concentrations. To reduce costs, it is desirable to have a separate collection system for high strength brewery effluents destined for anaerobic treatment. If complete treatment of all waste is desired, the low strength effluents should be treated aerobically along with the effluent from the anaerobic system.

Sludge Treatment, Disposal, and Utilization

One of the more expensive steps in wastewater treatment is the handling and dewatering of excess sludge. This is especially true for aerobic treatment systems where large quantities of sludge are produced. A high level of concentration of solids is required to reduce transportation costs or to keep fuel costs down if sludge is dried or incinerated. Landfill requirements sometimes also dictate maximum allowable moisture levels.

Excess sludge that is collected from settling tanks generally contains one to two percent solids. This sludge can be held in consolidation tanks for an additional two to three days to increase the solids concentration to about four percent. Solids can also be concentrated to about four percent by means of dissolved air flotation or by means of centrifuges. Sludge squeezed from the plastic sponge like media has solids concentrations of up to six percent.

Further dewatering may be done by means of pressure horizontal belt filters, by rotary vacuum filters, or by plate and frame or recessed plate filter presses. A pre-coat of fly ash or diatomaceous earth is often used in filter presses to prevent cloth blinding. Overall dewatering rates may be enhanced considerably by using chemical coagulants such as alum, lime, ferric chloride, and body feed such as fly ash, diatomaceous earth, or paper fiber.

Lime contributes to the value of the sludge if the product is to be utilized as fertilizer. Fly ash and diatomaceous earth are often brewery by-products that need to be disposed of anyway, whereas ferric chloride may sometimes be obtained inexpensively as pickle liquor, which is a by-product from the metal plating industry.

By adding about twenty-five percent lime, ten percent ferric chloride, and ten percent fly ash based on sludge solids, sludge from an aerobic treatment plant can be consistently dewatered to thirty-five to fifty percent solids in a filter press. Sludge from the deep shaft process, which has three to five percent solids after flotation, can be dewatered to sixteen to seventeen percent solids by using a belt press. Sludge from the breweries can be transported to landfill or the product to be useful as fertilizer.

Land Application of Brewery Effluents

A number of large breweries use the system of land application of selected high strength brewery effluents. The land application system has been combined with a year-round lawn operation to assure intensive management and utilization of nutrients.

The high strength segregated waste streams are collected in above-ground steel tanks with a combined hold-up capacity of three to four days. The tanks must be aerated to minimize odor problems. The pH is adjusted to an appropriate level for the particular soil. The effluents have to be passed through a screen to prevent coarse suspended matter from plugging up the rotating irrigation systems. The soil requires extensive preparation to be able to maintain a minimum water table of 1 m over the entire area. Turf grass may be harvested two to three times a year.

Concept of Zero Discharge Effluent

In a wastewater treatment facility, zero discharge theoretically means no discharge of any kind of pollutants into the any water downstream. But this is practically impossible and, the term 'zero discharge' is loosely used to define no liquid discharge into the environment. So, quite often, zero discharge and zero liquid discharge are used to mean the same thing. For all practical purposes, the concept of zero discharge necessarily means the following:

1. Recovery of reusable water/other materials from wastewater
2. Minimization or, no discharge of polluting substances into the environment away from the wastewater treatment facility.

In India, the concept of zero discharge essentially emerged from the situation where industry is unable to meet the discharge norms set by the government. This led to pollution of the environment and subsequent litigation.

Waste management has become a focus area in order to maximize recycling, reduce net waste and to achieve more sustainable operations.

Objectives:

1. The processes utilized for wastewater treatment do not generate any additional pollutants
2. Production of waste is minimized by suitable selection of unit processes and adjusting operating parameters
3. As far as possible, pollutants in the wastewater are transferred to solid phase (sludge)
4. Sludge is stored in a secured landfill
5. Recovery of reusable materials, especially water, is achieved

Since the concept of zero discharge system is to ensure essentially no discharge of pollutants into the environment, recovery of water gains primary importance. It achieves two purposes:

1. By reusing process water, utilization of natural water resources is minimized
2. Reuse of recovered water enhances the capacity of the brewery to efficiently utilize available water as well as control its quality to the required level

Design Considerations for Treatment Plant:

1. Quality of the wastewater to be treated
2. Efficiency of the treatment system
3. Ability of the treatment system to withstand variability in the quality of wastewater being treated over short-time (*shocks*) and long-term basis
4. Performance degradation of the machinery over a period of time
5. Operation and maintenance issues such as backwash and cleaning operations
6. Mass balance under different perceived operating conditions
7. Treatment of the treated effluent

To achieve the above-mentioned objectives, it is suggested that one should incorporate a reservoir for collection of treated effluent and filtration system after the existing wastewater treatment stream.

Filtration:

This is meant for the removal of suspended solids, color, and odor. Pressure Sand Filter (PSF) and Activated Carbon Filter (ACF) are normally utilized. Some of the design aspects that require attention are:

1. Design of PSF and ACF should also consider reducing the frequency of backwash to ensure that the performance of the zero discharge system is not getting affected

2. Treating the effluent with two parallel streams of filters and/or automated switching of filters, and provision for additional stand-by filters, are some of the options to ensure better filter performance

3. Removal of suspended solids should be achieved gradually (*viz.* larger particles are removed first and smaller ones in the next stage) by cascaded and suitably designed filters to ensure that shocks do not significantly affect the performance of filters

4. Under situations where color and odor removal by the filtration equipment is inadequate, ozonation prior to filtration may be considered

In the strict sense of zero discharge, wherein materials are brought into contact with beer, they require special safety measures (regulations)—especially in terms of microbiological safety. It must be noted that there is also a psychological barrier of recycling wastewater after treatment in brewing process. This is an obstruction to zero discharge.

In order to achieve zero discharge, we need to focus on partial recycling of wastewater after treatment for process. Then, it is advisable to incorporate an RO Plant so as to get water up to a level similar than the fresh water currently used in the brewery.

Issues:

Before implementing a zero discharge system, setting up a pilot plant with important unit processes is essential to ascertain that the chosen treatment procedure is able to successfully achieve its purpose—(i.e.) to test process validity and scalability, reducing energy consumption through optimal measures to reduce operating costs for successful operation and efficient maintenance of machinery.

1. It may be necessary to treat the wastewater resulting from the backwash of PSF and ACF in a sand-bed for the removal of suspended solids—and then pass this wastewater to the equalization tank to prevent frequent clogging of the filters.

2. The expected life-time of the filtering materials and their quantity have to be known in advance, as these solid wastes are to be transferred to the secured landfill, whose design requires the estimated waste solids output from each unit process.

3. In situations where color and odor removal by the filtration equipment is inadequate, ozonation prior to filtration may be considered. However, ozonation also produces hydroxides of transition metals, especially iron, that may be present in the wastewater. This will increase turbidity of the feed to the filters and may increase backwash frequency. However, there are certain benefits associated with this process.

4. After ozonation, residual ozone in the effluent should be removed. If this is not done, the machinery in the next treatment stage, and microbial growth in the biological digesters shall get affected.

5. Since process stability is very much important for any zero discharge facility, re-circulation should be considered during plant design whenever found appropriate and necessary.

Data Management

There are both risks and opportunities in water and wastewater management. Making informed business decisions to minimize risk and maximize opportunity requires effective data management.

Effective Data Management System

This covers best practices in data management, from establishing a data collection routine and ensuring the data is accurate, to creating key performance indicators and setting goals.

Data Collection

Successful data management enables cost-effective decisions to be made. Data management often goes beyond collecting usage and cost data from a monthly utility invoice. It includes identifying process areas, support functions, and facility operations that have the greatest opportunities for improvement. Strategies include tracking water metrics as part of process improvement activities, as well as installing water meters on processes that use large amounts of water and have a history of inconsistency. Understanding water use is critical to starting an effective conservation program.

Where to start?

1. Where is the water going?
2. How much water is used? What are the typical values?

3. How much water is discharged?

4. What is in the water discharged? What are the typical values?

5. How can water use and discharge be managed?

Using Water Meters to Identify Opportunities

Installing sub-meters at key locations is the best way to quantify and segregate water usage. Pulse output mechanical meters allow for automatic data collection, reducing measuring errors, and eliminating manual reading of the meters.

A single meter isn't capable of providing enough details on water usage in different process steps, so installing additional meters is highly recommended. If installing sub-meters is not possible, there are several other ways of estimating water volume. In places where water is transported in a constant flow, read the pump capacity and multiply this flow by the operating hours. One has to be careful when using this method, and not assume that the equipment is actually doing what it says on the nameplate. It is important to check the numbers against expected outputs.

Creating a formal mass balance of water and wastewater in the brewery is often costly and resource intensive. However, there are some data management steps that can be completed early in a water conservation program using a survey checklist.

Survey Checklist Example

1. Map the brewery's water distribution network and mark the routes of major pipes and drains on the site plan. Are the drawings up-to-date?

2. Identify the major points at which water is used.

3. Identify the major of wastewater discharge.

4. Identify the content of the effluent (yeast, trub, and so on), if possible.

5. Estimate the amount of water used and discharged at each major point.

6. Identify the water quality and availability at each [major] point.

7. Include designations for hot, cold and drainage systems.

8. Check water use in different areas of the brewery when production has ceased. If liquid is flowing through pipes or drains, either there is a leak or equipment has been left switched on (potential energy savings).

9. Label pipe work, valves and manholes for easy identification.

This task and checklist is often helpful for a new employee to fully and quickly understand the brewery and its processes. These checklists are also a good opportunity to partner with interns from a local university or trade program.

Keep in mind that a detailed water balance can be difficult to do because of evaporation losses. Evaporation, particularly from refrigeration plants, can account for as much as twenty-five percent of incoming water usage. Wastewater treatment also has a high rate of evaporation.

Flow Measurement Considerations

- How accurate does the data have to be?
- Does the data need to be trended or will a one-off measurement suffice?
- What is the size and material of the pipe?
- What is the operating pressure and temperature?
- What is the expected flow range (minimum to maximum)?
- Are there any existing meters that can be connected to a data logger or transmitter?
- When a tank is filled on demand, which is based on low-level/high-level switches, count the number of filling cycles. Determine the time it takes to fill one gallon in a bucket. This can help estimate the water flow.

Usage and Reduction Best Practices

Increasing yield and reducing beer loss should always be the first priority in any resource efficiency program. Reducing the amount of beer being spilled and wasted saves water, energy and ultimately, provides an immediate cost return.

Water reduction programs usually follow beer loss programs. There are usually some quick fixes for brewers just starting water reduction programs. The costs of even minor leakage is often overlooked or underestimated.

Water reduction projects have been difficult to justify, based on the cost of water; however, if the full cost of water is calculated, some projects may become more attractive, especially if future price increases are taken into account and value is put on business continuity and water reliability.

It is important to challenge the status quo ('This is the way it has always been done') when looking for water and wastewater reductions. The following questions can be useful when starting a water reduction initiative or when reviewing a mature program on a regular basis:

- Is the process or activity necessary?
- Is it necessary to use water?
- Why does the process use so much water?
- Can the amount of water be reduced?
- Can lower quality water be used?
- Can water be recovered elsewhere?

- Is the process authorized and legal?

- Is it necessary to produce wastewater or effluent?

- Is clean water going down the drain?

- Is the discharge authorized and legal?

- Would it be cost-effective to treat wastewater or effluent on-site for reuse?

Ways to reduce water use range from simple strategies, such as adjusting flow or installing water-conserving equipment, to more involved options, such as reusing water or switching to a low or waterless process.

Impact and Difficulty Matrix

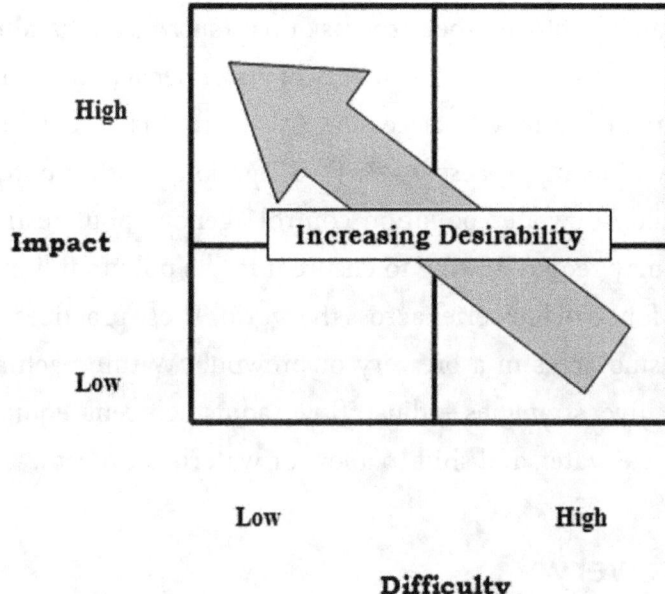

There are five general types of water-saving strategies, ranging from easy to more difficult implementation but also moving towards greater impact. One should consider these strategies when brainstorming ideas in cross-functional team meetings and other improvement efforts.

Key things to consider when starting water minimization efforts:

- Prioritize opportunities—what is most important for the brewery in the short term? What opportunities are good for the brewery?

- Consider water efficiency improvements in the context of other process improvements and lean performance goals to get the best results.

- Evaluate how process changes might affect wastewater volume or quality, or the impact on the environment (for example, switching from a water-based lubricant to an oil-based lubricant or solvent could have implications for workers' health and the environment).

- Consider which water efficiency best management practices and technologies make sense for the facility.

- Adopt visual controls or 'mistake proof' devices on equipment (for example, automatic shutoffs), and/or procedures to help ensure that process changes are effective and can be easily maintained.

- After testing potential solutions, make changes and evaluate actual performance and/or procedures to help ensure that process changes are effective and can be easily maintained.

- After testing potential solutions, making changes, and evaluating actual performance, be sure to develop or update the standard work for the activity so that employees can easily identify the best current way to perform an activity.

When evaluating water reuse opportunities, it is important to consider both quality and quantity. Not all processes need the cleanest, highest quality water. Many processes can reuse the 'waste' water from a process as an input to another process or use (such as air-handling condensate, reverse osmosis reject water, and so on), provided the quality of water needed for its intended use is matched. Testing and additional treatment may be necessary to ensure it is acceptable for future use. When evaluating the feasibility of using process water for irrigation or other outdoor uses, one needs to check with the local utility or water pollution control agency about restrictions on water reuse applications. The water may require testing to ensure it meets pollutant limits.

Best practices can be found in breweries across the world. These practices are presented according to the different water-using areas in a brewery or brewpub. Within each area, best practices are organized according to five strategies—adjust flow, adjust current equipment, change to new equipment, reuse or recycle water, and shift to a low or waterless process.

Best Practices: Brewery

Best Practices: Mash Cooker, Lauter Tun, Wort Kettle and Whirlpool

The following best practices can minimize water use, effluent flow and effluent strength:

- Do not fill the mash or Lauter Tun until it is too full. Train staff to add the correct amount of liquor and investigate the costs and possibilities of installing a meter to measure the volume of liquor being added.

- If new batches are frequently brewed, store surplus of wort and add it to the next brew.

- Store residual wort with trub for possible sale to farmers as animal feed supplement.

- Do not mix residual wort with surplus yeast—the mixture will start fermenting and the value of both waste streams will be reduced. Fermentable matter needs to be kept separate to maintain its value as an animal feed and yeast needs to be kept separate to maintain its value for food manufacturers.

- Separated grains and discharges to sewer will have a high BOD concentration. Excess settleable solids to sewer can cause blockage of pipes and accumulate at manholes. Where possible, use dry methods (brush or rake) to remove grains from the mash tun. There is no need to use water jets and subsequently discharge large amounts of effluent to the drain. Fit fine mesh baskets in the floor drains to collect and prevent grains from entering the drainage system.

Best Practices: Heat Exchangers

Compact heat exchangers are used in almost all breweries to recover heat from hot wort. The recovered energy can be used to preheat subsequent mash water or for washing purposes.

Since fermentation temperatures and cold liquor temperatures may vary among the different brews, automatic temperature control will allow for flow optimization of wort and cold liquor and will minimize water use.

Make sure the heat exchangers are well-maintained and regularly check the meter readings of the water flow. Pollution of the heat exchanger will negatively affect the heat transfer and cause an excess of water flow.

Check to ensure that cold liquor is not excessively warmed by ambient conditions.

It is important to check the heat exchanger capacities and the settings of the top-up valves, to prevent hot liquor tank overflows.

Best Practices: Fermentation Vessels

Single pass cooling of fermenters uses vast amounts of water. It is generally found at sites with inexpensive and readily available water (borehole or river extraction) and permission to discharge the cooling water to another location other than the sewer.

Closed loop systems will have pumps and control systems in place to regulate the cooling water flow through the fermenters cooling system. Make sure the pump size is adequate to cope with the maximum flow of cooling water when all fermenters are in use. Also, prevent overflow by setting the top-up level in the chilled water tank so that it is not topped-up until the reception tank is full. Install frequency controllers on the pump to fine-tune the water flow based on cooling needs. This will help to minimize the water flow as well as reduce energy use. Ensure procedures are in place to stop the cooling water supply when the fermentation process comes to an end.

Best Practices: Yeast Disposal

When the beer is drawn off, yeast slurry remains for removal. Avoid disposing this slurry to the drain, as it has a high BOD value and high suspended solids content. Also, large quantities of yeast lead to organic acids formation, affecting the pH (making the effluent more acidic).

An alternative may be to pass the slurry through a filter press or centrifuge to recover residual beer that may be reused in the process. The remains may be disposed separately or sold as animal feed additives. Yeast contains over forty percent protein and can be suitable as an animal feed supplement. If no filtration is possible or the residual beer cannot be reused, the complete slurry may be sold as animal feed and the liquid waste can be disposed to the drain.

Best Practices: Filtration

Diatomaceous earth, plate and frame, or rotary filters have traditionally been used to filter the beer prior to packaging; however, water consumption is high with these technologies. Alternatives include cross-flow or membrane filtration. Cross-flow filtration involves circulating the beer though a microfiltration cartridge containing a ceramic membrane. Yeast, bacteria, and other solids are retained on the membrane. This produces thick yeast slurry that can be disposed as described above. Since all bacteria are removed, no further pasteurization is needed.

Best Practices: CIP system

Using a CIP system is generally more efficient than manual cleaning. The advantages include:
- Increased vessel cleanliness due to chemicals and high temperatures employed
- High level of automation possible
- Reduced water and chemical consumption CIP is not a novel technology, yet it is often considered as such. There is significant opportunity to improve CIP, which offers water, cost and environmental savings

If CIP is not possible, a high-pressure hose will use much less water than a standard hose.

Best Practices: Chase Water

When beer is transferred in pipes, the pipes have to be cleaned and rinsed often. The operator needs to decide, in the case where pipes are rinsed with water, when the cleaning is ready and when the pipes can be filled with beer again. This process relies heavily on the judgment of the operator, which can lead to more water use than necessary. Automated interface detection systems may help, but tend to be unreliable. Purging with CO_2 is an option, but not often done because of the pressure involved and leaking of CO_2.

Another method is using a 'pigging' system. This is an engineered plug or ball that fits inside the pipe and is pushed through the pipe either mechanically or hydraulically to clear material ahead of the 'pig'. This can only be used where bends have a long radius and valves have bore openings.

Best Practices: Vacuum Pumps

Many vacuum pumps use water for cooling and for forming 'the seal' (a liquid ring). Instead of using water on a once-through basis, it can be recovered for reuse by recalculating the seal water via chillers or cooling tower. If a closed loop system is not possible, consider using the water as rinse water in the cask or bottle washer.

Best Practices: Good Process Design

It is good practice to design equipment with fewer parts and no points where fluid accumulates and that detergent cannot reach. This will reduce cleaning time as well as save water, chemicals and energy. Also, the design of pipes and drains can influence the accumulation of solids, making them more difficult to clean.

Best Practices: Glass Bottle Washing and Pasteurizing

Much of the waste from bottle washers and pasteurizers is due to overfilling of the feed tanks at the base of the units. This can be caused by leaks or by faulty valves or simply by an excessive top-up rate. In many washers or pasteurizers, the overflow points cannot be seen by the operators and overfilling of the feed tanks goes without notice. Ensure overflow points are visible for operators by extending the pipe to a position where operators can see it. Water metering will also allow identification of water use during periods the machines are not operating.

Consider the reuse of the final rinse water of the washers for the pre-rinse stage (or any other stage or application)—for single-use bottles, the rinse water used before the filling can be reused for many applications; however, as it may contain glass fragments, it should not be reused if there is a risk of contaminating the product.

Optimize the caustic dosing to the minimum quality standards to allow minimum water use during rinsing. Finally, it is a good idea to inspect the valves and the pipes of the washers and the pasteurizers regularly to detect leaks.

Best Practices: Cooling Towers

Evaporative cooling is a common and efficient way of dissipating thermal loads. Cooling towers and evaporative condensers require significant quantities of 'make-up' water to compensate for losses associated with evaporation, drift (or mist) and blow down (or purge).

A key parameter used to evaluate cooling tower operation is 'cycles of concentration' (sometimes referred to as cycles or concentration ratio). This is calculated as the ratio of the concentration of dissolved solids (or conductivity) in the blow down water compared to the make-up water.

Best Practices: Compressors

Refrigeration compressors often need cooling water. Since they produce excessive noise, these compressors tend to be isolated and inspected only when needed.

If possible, consider changing to a closed loop system with a cooling tower, or integrate the compressor cooling with another chilled water loop, like fermenting cooling. This chilled water loop is particularly effective when the brewery has several water-cooled units. A small bleed will be needed for hygiene reasons.

If a closed loop system is not possible, there may be a potential to reuse the water for various washing operations described earlier, like a cask washer or the CIP system. In this case, connect a solenoid valve to the cooling water supply to automatically cut off the supply when the compressor stops. Also install a frequency control to the pump of the cooling water supply to prevent manual tampering of the flow.

Best Practices: Steam Generation (Boiler)

Boilers and steam generators consume varying amounts of water depending on the size of the system, the amount of steam used, and the amount of condensate returned.

The key to operating an efficient steam boiler is to maximize steam generation and minimize losses to effluent by:

- Inspecting the boiler, condensate system and steam traps to find and promptly repair leaks
- Properly insulating steam and condensate pipes and hot well to decrease steam requirements and heat loss
- Minimizing blow down volumes by ensuring water treatment is optimized and the blow down automated
- Ensuring condensate return is maximized and the system is working effectively; recovering condensate for reuse will reduce water use, chemical use and energy consumption

Best Practices: Parking Lot/Landscape

Use of water for outdoor activities, such as cleaning up or landscaping can have a significant impact on costs. Most breweries pay sewer charges based on a percentage of incoming water usage. Excessive use of water for outdoor activities will result in higher water and sewer charges. This can be alleviated by negotiating the percentage with the authority or by installing a separate meter on water used for outdoor purposes.

Landscaping should be planned with drought-resistant and native plants where possible. Native plants require less water, and are better adapted to existing soil, climate, and wildlife. Cluster plants by similar irrigation needs.

Brewing

A water survey helps to identify the main users of water in the brewery. This survey can work as a starting point for further identification of possible water minimization measures.

By using water and effluent volume together with any sub-metering data, a water balance can be developed.

A water balance helps to:

- Understand and manage water and effluent efficiently
- Identify areas with greatest cost saving potential
- Detect leaks

Based on the water balance, a leak detection checklist can be used on a regular basis to determine possible 'hot spots' or areas where leaks could occur. A water balance should be reviewed on at least a twelve-month basis to ensure all changes and adjustments to the process and equipment are covered and the balance is up-to-date.

Leak Detection Checklist

A systematic program of leak detection and repair can prevent future water waste. Check the following areas thoroughly on a regular basis:

- Restrooms and shower facilities (in tank-type toilets, conduct dye tests to locate hidden leaks)
- Canteen, dishwashing facilities and food preparation area
- Wash-down areas and janitor closets
- Water fountains
- Water lines and water delivery devices
- Process plumbing, including tank overflow valves
- Landscape irrigation systems

Increasing employee awareness will ensure that measures taken to minimize the water use are understood and accepted by the operators and other staff. It will enhance the generation of ideas and it will make people proud when they can contribute to a more efficient brewery.

The following initiatives could be rolled out for an employee awareness program:

- *A water ambassador or champion*: He/she is the main contact for all water-saving projects, measures and metering. All employees know this person and know where to go to in case of questions or ideas. The water ambassador will also be responsible for the regular leak surveys.
- *An incentive program*: Employees are challenged to bring in ideas for water- or effluent-saving measures. For example, every three months, the employee with the best idea (highest water-saving potential, or most simple idea) is rewarded.

- *Employee education*: Set up a toolbox meeting on a regular basis to explain to, discuss with, and educate the employees on new findings, data and ideas on water savings. This will enhance the involvement and acceptance of any process adjustments, work procedures or equipment changes.
- *Alignment with home usage*: This will help the employee understand the importance of water minimization, and help to put things in perspective.

It is difficult to estimate the possibilities to save water with the introduction of best practices, since each brewery may be different from the next. However, there are some generally accepted ranges of reductions as shown in the table below:

Typical Reductions in Water Use

Water-saving measure	Possible application	Typical reduction in process use (%)
Closed loop recycle	Fermenter cooling	>90
Cleaning in place (CIP)	New CIP set	60
Reuse of wash water	PVC crates washing	50
Countercurrent rinsing	CIP set	40
Good housekeeping	Hose pipes	30
Cleaning in place	Optimization of CIP set	30
Spray/jet upgrades	Machines cleaning	20
Brushes/squeegees	Fermenter cleaning	20
Automatic shut-off	Pump cooling water	15

Conclusion:

A step-by-step study of our water cycle helps to significantly reduce water consumption. Making the water balance is useful to detect unknown water consumption, and is the basis for further optimization.

More practical common sense methods are developed to help reduce water consumption in beer manufacturing activities. This is known as the 'Water Scan Technique'—it is based on theoretical framework but starts from a water balance of the brewery, upon which ideas on how to manage the water system more efficiently, with concentration on reducing costs, are gathered. Good housekeeping and opportunities for simple direct reuse should be the main attention points at this stage.

Finally, the remaining wastewater can be including in an overall process scheme aimed at zero discharge of wastewater. In practice, hardly any process is operated hundred percent continuously, so an influent with stable characteristics is not always possible.

Application of new techniques for water/wastewater treatment or investment in less water consumption technologies must be considered. Opportunities for water-saving practices need to be classified based on their financial and technical consequences. Taking available resources into account, the most promising projects should be chosen and necessary planning for implementation of the proposal.

Community Benefits

Economic:

- Helps sustain community growth and business investment
- Results in better bond ratings that help communities in need of financing
- Helps cities and communities showcase their waterfront areas and commitment to clean water, thereby supporting new development and encouraging related commerce

Environmental:

- Helps decrease pollution in waterways, thereby reducing harm to the wildlife and ecosystem
- Reduces water and energy usage, leading to a decrease in greenhouse gas emissions, and puts less strain on natural resources
- Ensures that the community's natural resources and wildlife are protected

In addition to human needs, protection of endangered species and ecosystems will compete with available water supply. It is likely that the Government of India will introduce new laws and regulations that keep water 'in-stream' for species protection. The ongoing debate of water for human consumption, economic development, species protection, recreation, tourism and flood control will continue into the future.

Excessive water pollution can impact ecosystems. The high organic nature of brewery wastewater causes oxygen in a surface water to be depleted at a rapid rate, which negatively impacts living species and biodiversity. A number of water bodies in the United States remain above pollution levels considered safe for ecosystems. Additional regulatory restrictions are expected in the near future to address this problem.

16

Utilities Engineering

Introduction

The following six utilities are common to all breweries:

1. Steam
2. Water
3. Refrigeration
4. Electricity
5. Compressed air
6. Carbon dioxide

With the exception of CO_2 recovery and water treatment, utilities represent a source of energy supplied as needed in the brewing process and packaging operations.

The trend in new brewery projects, as well as in the expansion of existing breweries, is towards very large units in order to take advantage of the economics of scale in construction and production. The demand for utilities increases with the size and scope of the overall brewing process. Large scale in utilities follows large scale in brewing increasing in complexity with the demand for energy conservation and environmental pollution abatement.

Brewery capacity, location and arrangement vary among competing breweries in the industry. However, the brewing processes, packaging operations and utilities supply are similar. Expansion in brewery capacity can be realized by process changes and equipment additions. Unit costs for utilities will be reduced if the expansion is properly planned.

Steam

The boiler is the heart of the steam-generating plant. Water is evaporated in the drum of the boiler and distributed through piping to the users at pressure ranging from 7–15 kg/cm². Average distribution is 8.78–10.54 kg/cm².

At a point of use, the steam pressure is lowered through pressure reducing valve (PRV) to 2.8–3.2 kg/cm² and passed into the coils or jackets that provide the hot surface for heating process liquids. The steam condenses to water into the coils or jacket and condensate is returned to the boiler for reuse as boiler feedwater (recovery of condensate should be greater than eight percent). Condensate is pumped into a deaerating feedwater tank, make up water added as required, and steam injected to raise this temperature close to boiling. The water is then pumped into the boiler drum automatically to maintain a constant water level as loads vary in the system. Simultaneously, automated controls regulate the flow of fuel and air to the burner.

Boilers

There are two common types of boilers:

a. Water tube   : Water inside tubes is heated by fuel burning in a furnace that is surrounded by the tubes.

b. Fire tube   : Water in a boiler drum is heated by fuel burning inside tubes that is submerged in the water.

Furnace oil or natural gas are typically used fuels. With proper operation, overall steam-generating efficiency will be eighty to eight-five percent. Stack emission will be clean and non-polluting with low sulfur fuels.

Fuel at about 90°C is fed into the burner with air from a forced draft fan providing oxygen for combustion; ignition is started by an electric spark. The burning fuel is blown to the rear of the furnace.

Sufficient air must be introduced to burn a maximum amount of the fuel to CO_2. An excess of air is required to insure complete combustion, but too much will decrease the boiler efficiency by heating excess air not used in combustion. The percentage of CO_2 in the stack gas is a measure of efficiency. High CO_2 content means low excess air, high efficiency and high stack temperature. The temperature must remain high enough to prevent water vapor—another product of combustion—from condensing inside the stack.

Boiler Feedwater and Condensate Recovery

As heat is released in the mash kettle, wort kettle, bottle washer and pasteurizer percolator, the steam is condensed to hot water (condensate). Condensate is removed through a trap that will pass water but not steam, and forced to return to a condensate receiver by the pressure power pump unit (PPPU) to feedwater tank.

Condensate return can be high pressure at 3 kg/cm² or low pressure as discharged from a unit at 1 kg/cm². High- or low-pressure condensates are returned separately. Since not all steam is returned as condensate, make up water is added in the receiver (feedwater tank). Water is then pumped into a deaerator, steam is introduced to raise the water temperature and the gases vented out. This heated boiler feedwater is pumped into the boiler as required to maintain a constant level in the drum. The system is automatic and receiver and deaeration are sized large enough to accommodate fluctuating demands for steam in the brewery.

Fuels

Oil (furnace oil), natural gas, coal and biomass briquette are common fuels. Package boilers can be designed to use oil or gas by changing the burners. Coal and biomass briquette require additional handling equipment for unloading, storage, conveying and disposal of ash, as well as dust collectors to remove particulates entrained in the stack gas. The choice of the fuels depends upon the availability and price, both of which are frequently influenced by government policy, as related to the caloric value (CV) of each fuel.

Fuel	Average & Typical Range of Caloric Value
Natural Gas	8,900–9,400 kcals/m3
Furnace oil	10,667–11,000 kcals/kg
Coal	5,800 kcals/kg
Biomass Briquette	3,800 kcals/kg

Caloric values vary with the source of fuel. The burning of low sulfur fuels is required by law.

Energy-Efficient Boiler Plant

A *blow down economizer* was added to the boilers to reclaim heat from excess boiler water and is used to preheat feedwater for the boiler. By preheating the makeup water, the BTU/hr input of the boiler drops while the BTU/hr output remains the same.

The condensate of oxygen on the boilers is scrubbed before it returns to the boiler by a *deaerator*. This decreases the amount of boiler blow down and decreases makeup water, improving energy efficiency and requiring fewer chemicals to treat the make-up water.

Energy-Conserving Brewhouse

Process Heat: In the brewhouse, energy recovery reduces the need for steam. All of the heat for the brewhouse comes from steam supplied by two new, energy-efficient natural gas boilers. Heat lost from the boiling wort (unfermented beer) in the wort kettle is recovered and used to preheat future batches of wort with a *vapor condenser* on the exhaust stack. This translates into less energy demand, lower operational costs, and faster brewing time.

Besides reducing energy demand, there is another benefit to energy recovery—steam from the kettle condenses into water. In this way, the vapor from the cooking beer stays in the brewhouse and the area outside of the brewery does not 'smell like a brewery'.

Water

As a major raw material for the production of beer, getting water from a pure, fresh supply is obviously important. Availability of water suitable for brewing has always held a high priority in the list of the factors when selecting a brewery project.

Almost every water supply will require some treatment before use in the brewery. The original source may be rivers, underground wells, lakes and so on. Each will vary widely in chemical analysis and purity. Water analysis will determine the treatment required by the brewer. There are three general categories of water usage in the brewery:

1. Brewing
2. Boiler water (plus make up water)
3. Service, which includes all other usage

Water Treatment

Depending upon the original water source, all or some of the following operations may be required in water treatment:

- Aeration : Oxidation for removal of odors, colors
- Clarification : Addition of chemical for coagulation of suspended materials to cause settling prior to filtration or decantation
- Filtration : Removal of suspended solids or turbidity by sand or by diatomaceous earth for higher clarity; removal of oil
- Chlorination : Excess chlorine addition to oxidize organic materials, remove bacteria
- Carbon Filtration : Absorption by activated carbon for removal of odors, gases, organic chemicals, color
- Ozonation : Oxidation of organic impurities, bacteria
- Demineralization : By cation and anion exchange resins for removal of mineral salts in waters with high total dissolved solids content
- Rechlorination : Slight excess for bacteria removal

Treatment of boiler feedwater is usually needed to prolong the useful life of boilers, and steam distribution and condensate return system. Clean boiler interior surfaces are needed to ensure maximum efficiency of operation, rapid response to fluctuating steam loads and low operating cost. Feedwater enters the boiler drum to replace water that has been evaporated to steam. If the water in the boiler does not contain excessive dissolved solids the steam generated will be free of all impurities except dissolved gases. As steam is used, the concentration of the dissolved solids increases and is held at an acceptable level by intermitted or continuous removal through blow down of water from the drum.

Boiler Problem and Correction

Almost all raw water sources require special treatment for removal of dissolved solids and gases before being used as boiler feedwater. Impurities enter the system as make up water. Proper attention to boiler water chemistry will prevent major problems of corrosion and pitting of metallic surfaces, deposits of sludge and scale in boiler drums and tubes, foaming and carryover of impurities into the steam, and caustic embrittlement which causes crystallization and failure of metals.

Corrosion and Pitting

This is caused by the combined presence of dissolved oxygen and CO_2 in boiler water. Oxygen, liberated by heat, forms bubbles on metal surfaces. With CO_2 present, iron and oxygen react chemically. Iron is dissolved, leaving a pitted, structurally weakened metal. Iron and water also react with heat reaching a stable chemical equilibrium. Oxygen, when present, prevents equilibrium and continuous metal corrosion results. Carbon dioxide, also liberated by heat, leaves the boiler with steam causing corrosion in steam and condensate lines.

Corrosion and pitting are prevented by degasification in a deaerating feedwater heater. Condensate and make-up water are mixed in the condensate receiver (feedwater tank) and transferred to the heater by pump pressure. The heater operates at 100˚C and above. The solubility of dissolved gases decreases with temperature increase; they come out of solution and are vented to atmosphere. High-pressure steam from the boiler provides heat, and dissolved oxygen in the water is reduced to an acceptable level, 0.02 ppm.

Scaling and sludge deposits are controlled by treatment. Dissolved calcium and magnesium in the form of carbonates and sulfates cause hardness in water. When heated, insoluble carbonates are formed as scale in the drum and tubes. Scaling reduces tube diameter and slows heat transfer.

Foaming and carryover are prevented by maintaining solids content and alkalinity within proper limits. Well-designed baffles within the drums are helpful.

Caustic embrittlement is prevented by controlling alkalinity of the boiler water and by adding chemical inhibitors.

The steam from boilers so treated, obviously, cannot be used for direct injection into the product.

Refrigeration

A refrigeration plant is of central importance in a brewery. With about forty percent, refrigeration is one of the largest electrical power consumers and refrigeration directly influences beer quality. The final product is decisive; refrigeration is an essential tool. The right refrigeration capacity at the right time in the right place is essential.

In the brewery, refrigeration is used for cooling beer, chilled water for wort cooling, brine (propylene glycol or industrial alcohol) for attemperation and maintaining low temperatures in the cellars and hop storage rooms. Refrigeration is a process used for removing heat from a space or substance, to bring about a reduction in temperature, by transferring that heat to another substance.

Refrigeration Cycles

Most refrigerants undergo a series of evaporation, compression, condensation, throttling, and expansion processes, absorbing heat from a lower temperature reservoir and releasing it to a higher temperature reservoir in such a way that the final state is equal in all respects to the initial state. It is said to have undergone a closed refrigeration cycle. When air or gas undergoes a series of compression, heat release, throttling, expansion, and heat absorption processes, and its final state is not equal to its initial state, it is said to have undergone an open refrigeration cycle.

A. Compressor

A refrigeration compressor is the heart of a vapor compression refrigeration system. Its function is to raise the pressure of the refrigerant and provide the primary force to circulate the refrigerant. The refrigerant thus produces the refrigeration effect in the evaporator, condenses into liquid form in the condenser, and throttles to a lower pressure through the throttling device.

B. Condenser

The purpose of the condenser is to remove the amount of heat that is equal to the sum of the heat absorbed in the evaporator and the heat produced by compression. There are many different kinds of condenser.

C. Expansion Valve

The main purpose of the expansion valve is to ensure a sufficient pressure differential between the high and low-pressure sides of the plant. The simplest way of doing this is to use a capillary tube inserted between the condenser and evaporator.

D. Evaporation Systems

Depending on the application, various requirements are imposed on the evaporator. Evaporators are therefore made in a series of different versions. Evaporators and controls including space coolers for cellars and process cooling—beer, brine (propylene glycol, industrial alcohol, and so on.)

Unit of Refrigeration

The unit of refrigeration widely used in the industry is *ton of refrigeration*, or simply *ton*. 1 ton = 12,000 Btu/hr of heat removed—this equals the heat absorbed by one ton (2,000 lb) of ice melting at a temperature of 0°C over twenty-four hours or a ton of refrigeration is 12,000 BTU/hr, which is the amount of refrigeration required to make a one ton (2,000 lb) of ice in twenty-four hours.

Because the heat of fusion of ice at 0°C is 144 BTU/lb,

$$1 \text{ Ton} = \frac{1 \times 2000 \times 144}{24} = 12,000 \text{ BTU/hr}$$

Also, 1 ton = 3.516 kW

Refrigerant Safety

Refrigerant hazards stemming from leaks in the pipe joints, the rupture of system components, and the burning of escaping refrigerant depend on the type of refrigerant and refrigeration system.

Storage of Refrigerants

Refrigerants are usually stored in cylinders during transport and while on-site. During storage, the pressure of the liquid refrigerant must be periodically checked and adjusted. Excessive pressure may cause an explosion. Liquid refrigerants must not be stored above 54.4°C, although the containers are designed to withstand up to three times the saturated pressure at 54.4°C. If a container bursts, liquid refrigerant flashes into vapor. Such a sudden expansion in volume could cause a violent explosion inside a building, blasting out windows, walls and roofs.

Containers should never be located near heat sources without sufficient ventilation. They must also not be put in truck in direct sunlight. The valve of the container is attached by thread only. If the threads are damaged, the force of escaping vapor could propel the container like a rocket.

According to ASHRAE Standard 15-1994, in addition to the refrigerant charge in the system and receiver, refrigerant stored in the plant shall not exceed 150 kg. The receiver is a vessel used to store refrigerant after the condenser when necessary.

Refrigeration Plant

A refrigerating plant is an enclosure with tightly fitted doors to safely house compressors, refrigeration components, and other types of mechanical equipment. A refrigerating plant must be designed so that it is easily accessible, with adequate space for proper servicing, maintenance, and operation. A refrigerating plant must have doors that open outward and are self-closing if they open into the building; there must be an adequate number of doors to allow easy escape in case of emergency.

According to American Society of Heating, Refrigerating and Air Conditioning Engineers (ASHRAE) Standard 15-1994, a refrigeration plant must meet the following requirements:

- It must be ventilated to the outdoors by means of mechanical ventilation using power-driven fans or multiple-speed fans. Provisions for venting catastrophic leaks and component ruptures should be considered.

- There must be no open flames that use combustion air from the machinery room except matches, lighters, leak detectors, and similar devices.

A refrigerating plant of special requirements must meet the following specifications in addition to the general requirements:

a. Inside the room, there must be no flame-producing device or hot surface continuously operated at a surface temperature exceeding 427˚C.

b. There must be an exit door that opens directly to the outdoors or to a similar facility.

c. Mechanical ventilation for ammonia must be either continuously operating or equipped with a vapor detector that actuates a mechanical ventilation system automatically at a detection level not exceeding four percent by volume.

d. It must be provided with remote pilot control panel immediately outside the refrigeration plant to control and shut down the mechanical equipment in case of emergency.

Because the refrigerating plant itself is a fire compartment, the building structure and its material (including the door, wall, ceiling, and floor) should meet National Fire Protection codes. In a refrigeration plant, there are general and special requirements. The installation of mechanical ventilation and an oxygen or vapor sensor is mandatory.

CO_2 Gas Recovery

Introduction

The CO_2 recovery plant is an important factor that ensures optimum beer quality. Beyond that, CO_2 recovery also offers significant potential for cost savings and contributes to an ecologically responsible production.

Brewers are aware that how important the purity of the CO_2 is for the quality of the final beer. Even the least amount of residual oxygen in the CO_2 has a detrimental effect on the flavor stability of the beer.

The utilization of CO_2 from the own fermenters guarantees perfect quality—an advantage that no other source can offer.

Carbon dioxide in the brewery is generated by the yeast during fermentation, together with heat and alcohol. Because CO_2 is required at the end of the beer manufacturing process such as carbonation—(i.e.) to add the fizzy effect to the final beer, and purging process tanks—it reduces costs by recovering it during fermentation. Nevertheless, great care needs to be taken to avoid contamination of the final beer by air. Oxygen in final beer reduces the product shelf life and contributes to off tastes.

Additionally, a maximum CO_2 recovery yield is expected from this process. Today purities of around 99.998 percent can be achieved with the latest CO_2 recovery plants.

CO_2 from Fermentation

CO_2 gas is generated as a by-product of the breweries fermentation process. This is then collectively reclaimed from the fermentation area through adequately sized collection pipelines for common feed to the CO_2 gas recovery system. The gas at this point will be at low pressure and combined purity of > 98.5 percent.

CO_2 Collection and Recovery

Collection of CO_2 should start after twenty-four hours from the start of fermentation (initially CO_2 should be vented out) and stop collection when the gravity falls to 5.0–4.5 Plato.

The recovery plant compresses CO_2 gas, elevating the pressure to approximately 18 kg/cm² for CO_2 gas processing that being: washing, purifying, drying and CO_2 gas condensing.

Once compressed, CO_2 gas is treated for removal of impurities typical of these sources by high-pressure high-efficiency CO_2 gas washing (scrubbing) providing a CO_2 purity of min 99.9 percent.

The system further enhances the gas quality through proper CO_2 gas purifying. This is accomplished by an activated desiccant bed for gas drying to a dew point of -40°C at pressure followed by carbon polish filter, again subject to raw gas and process conditions. Once the operation is completed, the final gas will be odor free, color free and taste free, preparing for the last stages of purification.

As a means of final purification the CO_2 gas is condensed (separation of non-condensable gases). Carbon dioxide gas condensing is accomplished by use of an independent refrigeration system that liquefies CO_2 gas at approximately 18 kg/cm² and -24°C. The non-condensable gases present in the CO_2 gas are separated and purged from the system automatically and reused for regeneration gas within the plant.

Liquid CO_2 leaving the CO_2 condenser flows by gravity to a liquid CO_2 purification system to achieve a final liquid CO_2 purity of 99.998 percent. Thereafter, high quality liquid CO_2 is pumped to a liquid CO_2 storage tank for handling the liquid CO_2 for use for beer carbonation.

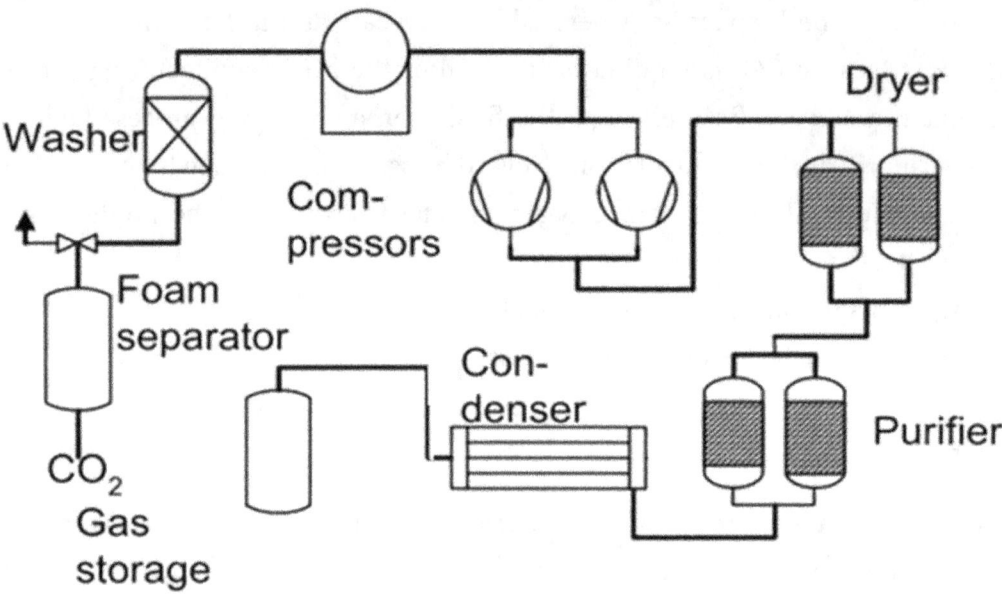

Figure 16.1: The Standard Method for CO$_2$ Recovery

Compressed Air

Compressed air systems are the most important utility in breweries and are often the most misunderstood. It is essential for the brewers and engineers to provide an understanding of common compression systems and operation techniques in breweries.

The interconnecting aspects of a compressed air system and to show advantageous solutions to problems that arise when planning and operating such systems properly. The planning of a compressed air system must be carried out carefully, using a comprehensive layout of information and criteria that can be clearly understood.

Usages of Compressed Air in the Brewery:

- To push fluids through piping and empty tanks, in the form of dry, oil-free, sterile air
- To aerate wort, yeast, or water, in the form of dry, oil-free, sterile air
- As an energy carrier for the pneumatic transportation of solids, such as spent grains, whole malt, sugar, and filter media, in the form of oil-free, and, where necessary, dry air
- As a purge gas to displace CO$_2$ from tanks prior to being cleaned-in-place with caustic, in the form of oil-free, sterile air
- To modulate valves in valve control operations, in the form of dry, oil-free air
- As an energy carrier to drive air tools, in the form of dry air
- Plant maintenance and control

17

Brewing Calculations

Introduction

The calculations applied to brewing materials, products, and processes for the purposes of quality, specifications, materials balance, and financial accounting—and in some cases, product identity and as a basis of taxation—require simple formulas and mathematical operations.

Beginning with brewing materials, and continuing on throughout the process to the 'finished' beer, there is a single, reliable 'reference standard' is identified as 'extract' plus associated adjectives that further describe its nature and derivation.

Extract in the brewing sense encompasses the total available soluble mass in a brewing material 'as is', and/or potentially through processing.

The potential extract of brewing materials is expressed as percentage by weight 'as is' and also on a 'dry basis' (kilo of extract per 100 kg of materials as is and mass units of extract per 100 units dry basis of materials). The extract values of worts and beers are expressed in terms of *degrees Plato* (weight percentage) and for computational convenience as kilograms per hectoliter (kg/hl). Plato values are acquired directly from standard Plato hydrometers, by means of specific gravity measurements as related to standard Plato tables, and approximately by calculation. The Plato is defined as the extract in percentage by weight of the wort, beer, or other solution.

For example, wort is determined to have a Plato value of 12.5°P from its specific gravity of 1.05048 and we required the kgs of extract contained in 1 hl of the wort. The solution is:

Wort specific gravity $\times$ °P = kg extract (solute)

This approximation remains viable because of the lack of a more appropriate reference better related to the variability in compositions of worts of different origins.

Balling's and Brix are not equivalent to degree Plato and are not compatible with the compositions of standard methods.

Balling and Brix hydrometers are readily available and preferred in microbreweries because they are less expensive. These instruments may be successfully used for quality control purposes provided one modulus is used universally; all instruments are practically standardized from time to time; and slight differences in inter-laboratory results are acceptable. However, a first alternative choice to replace the Plato hydrometers is a set of specific gravity hydrometers. The Plato values derived by the 'Lincoln equations' correspond identically to those of the ASBC tables.

I. The Lincoln Equations and Extensions

The Lincoln equations (1987) are applied to the calculations of wort and beer solute concentrations and specific gravity. The purpose of the Lincoln equations is the replacement of cumbersome mathematical tables with simple parametric equations easily incorporated into on-line computer programs and handled programmable calculator. Obviously, the equations can be directly substituted into computer computations for the parameters they represent, thereby simplifying the equation.

1. Kilograms of extract per hectoliter from degree Plato:

$$E_{kg/hl} = 0.9974 \times (P^{-1} - 0.00382)^{-1} \tag{1}$$

2. Specific gravity calculated from degrees Plato:

$$SG = 1 + P \times [258.6 - (0.8796 \times P)]^{-1} \tag{2}$$

3. Degrees Plato calculated from specific gravity:

$$°P = [463 - (205 \times SG)] \times (SG - 1) \tag{3}$$

Where:

$E_{kg/hl}$	= Kg of extract/hectoliter
E_{kg}	= Kg of extract
P	= Degree Plato (°P)
SG	= Specific gravity
V_{hl}	= Volume in hectoliter

II. Fundamental Descriptors of Brewing, Materials & Wort

A. Extract & Theoretical Yield

The percentage by weight of water-soluble extract derived from materials by means of standardized laboratory procedure such as EBC or ASBC is termed as the *theoretical yield.*

Two values are normally reported for malts—*fine grind* and *coarse grind extracts.* The coarse grind value more closely approximates average production results, but does not represent total available soluble (extract) approached by modern, highly efficient capabilities. The use of the fine grind value is necessary for production efficiency calculations.

B. Extract (Total) and Production Yield (Brewhouse, Mashing)

The percentage by weight of water-soluble extract (wort solute) derived from a single material (malt) or a combination of materials (malt and adjuncts) is based on the total wort delivered to the wort kettle (fill).

Total Kg of extract to wort kettle = hl Wort × Kg/hl Extract

$$\Sigma E = V_{hl} \times E_{kg/hl} \tag{4}$$

The total weight of extract yielding materials is:

$$\text{Total Kg of materials used} = M_1 + M_2 + M_3 + \ldots\ldots + M_n = \Sigma M \tag{5}$$

Where M represents the weights (kg) of the individual brewing materials of the formulation.

$$\text{Production Yield (\%)} = \frac{\Sigma E}{\Sigma M} \times 100 \tag{6}$$

C. Production Efficiency (Brewhouse, Mashing)

The total available extracts (ΣE_a) on the basis of material analysis for production efficiency (brewhouse) is:

$$\Sigma E_a = M_1 Y_1 + M_2 Y_2 + M_3 Y_3 + \ldots + M_n Y_n \tag{7}$$

Where:

  M = Weights of the individual materials
  Y = Analytical yields of the individual materials
  ΣE = Total weights of extract in wort

Production efficiency (Mashing) is:

$$\text{Efficiency of Mashing Operation (\%)} = \frac{\Sigma E}{\Sigma E_a} \times 100 \tag{8}$$

D. Additional Extract Added to Wort Kettle

The yield and efficiency of any extract-contributing materials added to wort kettle are calculated on the basis of observed increase(s) of wort kettle extract (ΔE) and wort volume (ΔV). It is to be remember that wort kettle volumes are usually measured at an elevated temperature (usually 95–100°C but possibly as low as 75°C) and the Plato values are measured at 20°C, thus requiring a proper volume adjustment, usually -4 percent wort expansion due to temperature (20°C versus 100°C).

The yield of kettle additive is:

$$\text{Kettle Additive (KA) Yield (\%)} = \frac{(\Sigma E_2 - \Sigma E)}{M_{KA}} \times 100 \qquad (9)$$

The efficiency of kettle additive is

$$\text{Efficiency KA (\%)} = \frac{(\Sigma E_2 - \Sigma E)}{M_{KA} \, Y_{KA}} \times 100 \qquad (10)$$

Where,

ΣE	= Total wort extract mass before additive addition
ΣE_2	= Total wort extract mass after additive addition
M_{KA}	= Mass of additive used
Y_{KA}	= Yield of kettle additive

E. Original Extract (OE) of Wort and Standard Reference Wort (SRW)

The total volume of wort as delivered to the wort kettle, containing the full complement of solutes, hopped, finished, boiling, as is, for transfer from the wort kettle, represents the *Standard Reference Wort (SRW)* for a given product. The composition of the SRW is used as reference for all subsequent quality and efficiency calculations as well as component ratio.

The Plato value, or total extract, of this standard reference wort is termed the *original extract of the wort* OE_w. The original extract of the beer in process and the final beer, OE_b, may or may not be identical or even of the same magnitude as that of the OE_w. Thus, during processing, the ratio of components to respective OE values becomes as important as concentration levels in signaling disproportionate changes of individual components. A very practical example is that of the hops IBU values, which are by necessity compared on the basis of an SRW of constant OE, as reference. Similarly, high-gravity wort can be related to their resulting beers only on this common basis.

An SRW of constant OE at transfer from the wort kettle is often achieved by process yielding wort in the wort kettle at the end of boiling and ready for transfer having an OE slightly higher than the SRW-designated at P at this point. The P is then adjusted by the addition of calculated required volume of hot (almost boiling) process water to yield the constant SRW extract (P) designated. This simple calculation is hl of process H_2O required to dilute wort from P_1 to P_{SRW}.

$$\Delta V = (V_{SRW} - V_1) \qquad (11)$$

Where:

$$V_{SRW} = \frac{V_1 \times SG_1 \times °P_1}{SG_{SRW} \times °P_{SRW}}$$

ΔV	= Required dilution H_2O volume to be added
V_1	= Volume of wort in wort kettle
V_{SRW}	= Volume of SRW wort
SG_1	= Specific gravity of wort in wort kettle
SG_{SRW}	= Specific gravity of SRW wort
$°P_1$	= Degree Plato of wort in wort kettle
$°P_{SRW}$	= Degree Plato of SRW wort

F. Effects of Yeast Addition

The wort loses its identity upon addition (pitching) of yeast and the onset fermentation. The wort at this point becomes fermenting beer. 'Pitching yeast' is usually in the form of a suspension of yeast cells in 'end-fermented beer'. In case of low yeast density, the beer may represent as much as fifty percent (w/w) and have a significant and misleading effect on any specific gravity measurement on the wort to which it is added. Mathematical corrections for the deviation can be calculated by means of the preceding formulas.

Krausen and blending practices of fermenting beers are amenable, though the calculations are more complex, to 'blending calculations' of the finished beer.

The continuity and reliability of the extract reference parameter throughout the continuing process will be developed in 'Beer Calculations'.

G. Fermentable Wort Extract, Non-Fermentable Wort Extract, and End Fermentation

The extract constituents are classified into two groups:

a. Fermentable species
b. Non-fermentable species

However, a third group is present, consisting of species only partially fermented to various degrees in normal fermentations (such as maltotriose). These three parameters are critical to wort and resultant beer description and quality. All three values are predictable on the basis of a suitable rapid fermentation wort analysis or carbohydrate analysis for the individual fermentable sugars. The rapid fermentation is preferred because of the simultaneous yeast performance insight afforded. The calculations require the original extract (or specific gravity) value of the wort and the specific gravity and real extract values of the resultant beer produced by the sample fermentation.

The fermentable extract of wort (% w/w) = Real degree of fermentation

$$FEw\% = RDFw = \frac{(OE_w - RE_b)}{OE_w} \times 100 \tag{12}$$

Where:

FE_w	= Fermentable extract of wort
RDF_w	= Real degree of fermentation of wort
OE_w	= Original extract of the wort
RE_b	= Real extract of the resultant beer

The non-fermentable wort extract is

$$\text{Non-fermentable } E_w\% = (100 - FE_w) \tag{13}$$

This is trivial and not usually reported. In case of the end-fermented beer, it represents the limits of fermentation under the test/production conditions and includes a negligible amount of trace-fermentable materials.

The end fermentation, FE_w, or end-fermentable extract of wort (yeast fermentable extract) is calculated by Eq. (13) employing data determined by the end-fermentation procedures.

The rapid fermentation procedure is usually designed to yield resulting laboratory beers identical to those produced by established standard production practices. End fermentations represent a very close approximation of the total fermentability of the wort.

H. Process Extract Profile (Materials, Wort and Beer)

A process extract profile includes a complete summary of the total extract throughout the process of a given beer formulation and process. The profile consists of the total extract values for the following parameters:

- Each brewing materials of the formulation
- The sum of the material extracts of the formulation
- The first wort (*note volumes of the wort in the wort kettle at beginning, middle, end*)
- Lautered wort (*collect approximately five samples at equally spaced volume increments, beginning to end*)
- End-of-lauter
- Extract remaining in spent grain

- Wort kettle full wort (*note immediately at fill point, time and volumes in the wort kettle*)
- Wort at the end of boil (*note volume of the wort in wort kettle, `P analyze wort for critical parameters such as IBU, color, etc.*)
- Wort delivered to fermenting vessel

Continue throughout the remaining process as beer, reporting OE_b and back-calculating all beer parameters to the base of the OE_{SRW}.

III. Beer Calculations

The yeast metabolizes the fermentable extract of the wort to varying degrees, yielding the principal product of ethanol (alcohol), carbon dioxide (CO_2), and increased biomass (yeast).

Wort + Yeast $\rightarrow$ Ethanol + $CO_2\uparrow$ + By-products +

Fermentable extract $\rightarrow$ Residual variable O_2 to designated%

Non-fermentable extract $\rightarrow$ virtually unchanged

For practical purposes, and with much usefulness, these impediments to precise calculations have been overcome, permitting simple observations and calculation to maintain high levels of quality control.

The specific gravity of a decarbonated beer and the corresponding Plato value are determined in the conventional manner. The Plato value so determined is termed as *apparent extract*. The apparent extract is of no precise physical significance, but is an extremely useful relative value universally employed and included in the most sophisticated beer specifications.

All calculations based on apparent extract values are similarly identified as being an apparent 'x' value and are approximate but useful data.

This resolves the impediment of the aqueous/ethanol solvent.

The second problem is maintaining the continuity of the total extract balance throughout the process. This problem was resolved through the research of Balling and Plato, who established a mathematical relationship between the mass of extract depleted in the production of the two principal reaction products, ethanol and carbon dioxide. The Balling-Plato equations consist of one set of precise, analytically based parameters and a second set of approximate equations, primarily based on calculated or approximate non-dimensional numbers. The Balling values are applied to the Gay-Lussac equation are:

Wort Original Extract	$\rightarrow$	Ethanol	+	$CO_2\uparrow$	+	New Yeast	(14)
grams: 2.0665		1		0.9565		0.11	

$$1.0665$$

The associated equation is:

$$OE = \frac{(A \times 2.0665 + RE_b)}{100 + (A \times 1.0665)} \times 100 \qquad (15)$$

Where:

OE_w = Original extract of the beer, calculated

A = Ethanol (alcohol), % w/w

RE_b = Real extract of the beer

Equation (16) enables the continuous calculation of efficiencies and materials balance throughout the process on the basis of beer analysis, observed volumes, and the accommodation of any blending or dilution of the original unit volume.

Point to remember: Wort and beer specifications (SRW and SRB) should include a *precise* OE value for both wort and beer.

The simplest unit references throughout the process are:

1. Total extract per SRW (total extract, kg per brew of standard reference worth of OE = n °P)
2. Total volume (hl) of equivalent standard reference worth (V_{SRW}) per brew
3. Total volume of a unit of standard reference beer (hl, specification to include precise OE_b)
4. 'Dock yield' terms in specific units (hl of SRB including precise OE_b)

A reliable and consistent means is accordingly established to relate the beer parameters with the original extract of wort, OE_{SRW}.

Precise Calculations

The term 'precise calculations' is applied in the sense that the calculations are based on precisely defined parameters and units whose values are acquired by standardized analytical methods.

By knowing the alcohol content as well as the real extract, it is now possible to determine the original gravity (OG) of the beer, which is the theoretical extract the beer had before fermentation. For this, the famous Balling formula is used:

The Balling formula is based on the empirical knowledge that during fermentation 2.0665 g extract is converted into:

- 1 g alcohol
- 0.9565 g CO_2
- 0.11 g yeast

Thus, the extract before fermentation is:

- Fermentable extract = A × 2.0665 (%, w/w)
- Non-fermentable extract = RE (%, w/w)

Calculation of the concentration of extract before fermentation:

- Fermentable extract = A × 2.0665
- Non-fermentable extract = RE
- Total = (A × 2.0665 + RE)

That concentration refers to beer. As we are interested in the theoretical concentration of the wort prior to fermentation, it is necessary to consider the losses during fermentation due to yeast and CO_2 formation. Therefore, the following mass balance is used:

- 100 g beer refers to:
- (100 g + yeast + CO_2) wort
- (100 g + A × 1.0665) wort

That is the reason for the denominator and completes the Balling formula.

For the original extract of the wort from the beer analysis (OE_b):

$$OE_b = \frac{(A \times 2.0665 + RE) \times 100}{100 + (A \times 1.0665)} \qquad (16)$$

The *apparent extract* is the measure is specific gravity of beer, which is a combination of the extract content and the produced alcohol. The alcohol causes a thinning effect giving a lower reading of the actual extract content, hence the terminology 'apparent'.

Apparent Degree of Fermentation is the difference between the extract in original wort and the apparent extract in the fermenting wort divided by original extract.

$$\text{Apparent Degree of Fermentation (ADFw) (\%)} = \frac{(OE_w - AE_b)}{OE_w} \times 100 \qquad (17)$$

Where:

OE_w = Original extract of the wort (°P)

AE_b = Apparent extract of the beer (°P)

The *real extract* is determined by removing the alcohol from beer sample by distillation and replacing the volume with distilled water. The *real degree of fermentation* is the difference between the extract in the original wort and the real extract in the sample, divided by the original extract.

$$\text{Real Degree of Fermentation (RDFw) (\%)} = \frac{(OE_w - RE_b)}{OE_w} \times 100 \qquad (18)$$

Where:

OE_w = Original extract of the wort (°P)

RE_b = Real extract of the beer (°P)

The original extract of the wort from which a beer was produced is back calculated on the basis of the beer analysis, ethanol (alcohol), and real extract. The value is described as the original extract or original gravity of the beer in Plato units. This is a critical calculation for product identity, quality control, mass balance (yield, efficiency), and economic purposes. It is obvious that this calculated original extract of the beer relates directly to that of the original kettle wort extract only in case where the wort and beer have not been diluted or where appropriate adjustment is made for such dilution.

Attenuation is the measure for the extract decline, and defined as the decline of specific gravity of wort during fermentation

For the real attenuation (RA):

$$RA_b = (OE_b - RE_b) \qquad (\text{\textit{All in} } °P) \qquad (19)$$

The real attenuation (RA) of a beer (RA_b) is the mass of extract in °P units, fermented to produce the given beer, based on the calculated OE of the beer.

For the real degree of fermentation of wort (RDF_w):

$$\text{RDFw (\%)} = \frac{(OE_w - RE_b)}{OE_w} \times 100 \qquad (20)$$

The real degree of fermentation of wort is the percentage of the total wort extract that is fermentable as determined analytically by a rapid fermentation procedure.

For the real degree of fermentation of beer (RDF_b):

$$RDFb\ (\%) = \frac{(OE_b - RE_b)}{OE_b} \times 100 \times \frac{1}{1 - (0.005161 \times RE_b)} \tag{21}$$

The real degree of fermentation of beer is the percentage of the calculated original extract of the beer that was fermented in process.

Also, by the approximation calculations:

$$RDFb\ (\%) = \frac{(OE_b - RE_b)}{OE_b} \times 100 \tag{22}$$

$$RDFb\ (\%) = \frac{RA_b}{OE_b} \times 100 \tag{23}$$

For the carbohydrate content of beer (approximate, in % weight):

$$\text{Carbohydrate per 100 g of beer} = RE_b - (\text{protein} + \text{ash}) \tag{24}$$

For the carbohydrate content of beer (approximate, in % weight/volume):

$$\text{Carbohydrate per volume of beer} = \frac{\text{Carbohydrate per 100 g}}{(V \times SG_b)} \times 100 \tag{25}$$

The calorie content of beer (ASBC approximate, not applicable for diet beer) is:

$$\text{Kcal per 100 g of beer} = 6.9\ A_b + 4\ (RE_b \times Ash_b) \tag{26}$$

Where the coefficients represent Kcal/g of the chemical species, and:
   A_b = Ethanol (alcohol)
   RE_b = Real extract all in grams/100 g of beer

The calorie content of beer (preferred, applicable to diet beers) is:

$$\text{Kcal per 100 g of beer} = 4N + 4C + 4G + 7A + 3OA \tag{27}$$

Where the coefficients represent Kcal/g of the chemical species, and:

N = % wt protein

C = % wt carbohydrate

G = % wt glycerine

A = % wt ethanol (alcohol)

OA = % wt organic acids

All in grams/100 g of beer.

The estimated freezing point of beer (Kolbach, 1940) is:

$$\Delta Fp = 0.42\,A + 0.04\,RE + 0.2 \qquad (28)$$

Where:

ΔFp = Freezing point depression (˚C)

A   = Ethanol (alcohol), % wt/wt

RE   = Real extract, % wt/wt

Knowledge of the freezing point of beer is particularly valuable for protection of the packaged beer shipped in subfreezing climates.

For the apparent molecular weight of real extract solids (applied to dealcoholized beer or dilute worts only) (Weissler, 1965). If the milliosmolality of the sample is known, then the simplified calculation is:

$$MW_b = \frac{(E \times 10^4)}{(\Phi)} \qquad (29)$$

Where:

E = Real extract of beer/wort ($^{\circ}$P)

Φ = Milliosmolality of the sample (*10^{-3} mole very convenient for small masses or concentrations*)

Worts must be diluted but this does not alter the validity of the calculations.

For the molal concentration of wort and beer solutes (apparent molality of wort and dealcoholized beer):

$$\text{Apparent molal concentration} = ^{\circ}P \times (100 - ^{\circ}P) \times MW_b \times 10^3 \qquad (30)$$

This equation is basic to all colligative property calculations.

The temperature of maximum density of beer (Weissler, 1965) is:

$$°C_{md} = 4 - (0.65\ RE_b - 0.24\ A) \tag{31}$$

Where:

$°C_{md}$ = Maximum density of beer

RE_b = Real extract of beer

A = Ethanol (alcohol), % wt/wt

Equation (31) is particularly useful in the placement and design of the tank attemperators and in the blending of beer at low temperatures (< 5°C).

The depression of the temperature of maximum density of water by wort solutes is given by:

$$(\Delta\ °C_{md})\ w = -\ K_{\Sigma}\ x\ \frac{°P}{(100 - °P)} \tag{32a}$$

The temperature of maximum density of wort is given by:

$$(°C_{md})_w = 4.0 + (\Delta\ °C_{md})_w \tag{32b}$$

Where:

$(\Delta\ °C_{md})_w$ = Depression of the temperature of maximum density of water (*Caution: a negative value*)

$(°C_{md})_w$ = Temperature of maximum density of wort

$-\ K_{\Sigma}$ = Combined constants = 5.435 for normal worts

$°P$ = Degree Plato of wort

The molal maximum density constant is desirably determined for each beer type—such as high and low attenuating worts, and worts of high levels of unusual constituents such as oats, wheat. This equation is useful in production calculations involving freeze-concentration of worts/beers or the so-called 'ice beer' processes.

The specific heats (SH_w) of worts are required for engineering and process performance calculations although the determination of specific heats well within the requirements of practical applications.

$$SH_w = (-4.5515) \times °P + 3.6749 \tag{33}$$

Equation (33) has been found to be reliable for practical application to normal worts over a range of 0.5–25 °P and for normal brewing liquid adjunct solutions within the range of concentrations.

Usually the specific heat of milled barley malt is expressed as its 'water equivalent'. The specific heat SH_m of milled barley malt is approximately 0.42. The water equivalent given as 1,000 kg malt is equivalent to 4.2 hl of water in heat content.

The purpose of the water equivalent is to facilitate heat balance computations in the brewing process. It states that the heat content of a given mass of malt may be expressed in a common denominator of a volume of water equivalence, thus simplifying the heat balance calculation. Specific application is calculating the required temperature of mash in water knowing the malt temperature and the desired final temperature of the combined water and malt in the mash kettle.

The water equivalent of other brewing materials may be similarly calculated employing:

$$W_{eq\,(A)} = \frac{1000 \times SH_A}{100} \tag{34}$$

$$W_{eq\,(malt)} = \frac{1000 \times 0.42}{100}$$

$$= 4.2 \text{ hl of water equivalent}$$

Where:

$W_{eq\,(A)}$	= Heat units of 1000 kg of materials 'A' equivalent to hl of water
SH_A	= Specific heat (heat content) substance 'A'
100	= k = Heat content of 1 hl of water = 100 kg water
1000	= k = Unit reference weight of 'A'

The precise and approximate equations for the ethanol (alcohol) percentage weight conversion to ethanol (alcohol) percentage volume are:

a. Precise equation:

$$A_{\%v} = \frac{A_{\%w} \times SG_b}{0.791} \tag{35a}$$

Where:

$A_{\%v}$	= Ethanol (alcohol), % volume
$A_{\%w}$	= Ethanol (alcohol), % weight
SG_b	= Specific gravity of beer

b. Approximate equation:

$$A_{\%v} = 1.25 \times A_{\%w} \qquad (35b)$$

c. Conversely, approximate conversion of ethanol (alcohol) percentage by volume to ethanol (alcohol) percentage by weight ($A_{\%v}$ to $A_{\%w}$):

$$A_{\%v} = 0.8 \times A_{\%w} \qquad (35c)$$

Equation (35c) is most convenient and often used for rapid estimates.

For ethanol (alcohol), percentage weight ratio to real extract percentage weight (Weissler, 1933) is:

$$(A_{\%w})_b \div (RE_{\%w})_b = \Psi \qquad (36)$$

Where:

$(A_{\%w})_b$ = Ethanol (alcohol), % weight of beer

$(RE_{\%w})_b$ = Real extract, % weight of beer

Ψ = Ratio, positive number

If:

Ψ is < 1.0; the beer is unbalanced with high body;

Ψ is = 1.0; the beer is balanced;

Ψ is > 1.0; the beer is unbalanced with light body;

This standard has become an important criterion for anticipating the relative satiating effects of beers.

The calcium ion to oxalate ion molar ratio of beer must be equal to or greater than 10 to 1 to insure gaseous stability (Brenner, 1958):

$$(Ca^{2+}) :: (C_2O_4^{2-}) \geq 10$$
$$(Ca_{ppm}) \geq 4.55 \, (C_2O_4)_{ppm} \qquad (37)$$

Brenner demonstrated that a major cause of gushing beer was post-precipitation of calcium oxalate. To avoid this problem, a brewer must provide sufficient calcium ion to satisfy the solubility product requirement for oxalate ion precipitation (removal) prior to packaging. For adequate assurance of oxalate removal, Brenner recommends that the molar ration of calcium ion should be greater than ten.

Equation (37) conveniently indicates that to meet this molar ratio, the calcium ion concentration in mg/l (ppm) must be 4.55 times, or greater than that of the oxalate ion concentration in mg/l (ppm). This condition is so important that an example calculation is in order.

Suppose a range of calcium ion as ppm in beers to be 35–50 ppm. Then, this range of Ca^{2+} offers effective protection for the $C_2O_4^{2-}$ as follows:

If Ca = 35 ppm, then
$(Ca_{ppm}) \geq 4.55 (C_2O_4)_{ppm}$
$35 = 4.55 (C_2O_4)_{ppm}$
$(C_2O_4)_{ppm} = 7.69 = 8$ maximum
If Ca = 50 ppm, then
$(Ca_{ppm}) \geq 4.55 (C_2O_4)_{ppm}$
$50 = 4.55 (C_2O_4)_{ppm}$
$(C_2O_4)_{ppm} = 10.99 = 11$ maximum

With the calcium ion concentration range of 35–50 ppm, the highest tolerable oxalate ion concentration is 11 ppm.

IV. Formulation Calculations

The requirements for the design and operation of a brewery on sound business and scientific principles require a written technical manual outlining:

- Product definition including specifications
- Process design
- Formulation requirement and constraints

In all cases, it is necessary to follow fundamental principles for new brewery, which are:

- Planning
- Operating brewery audit
- Operating brewery reorganization

The following equations are in sequence based on the information from these principles.

A. Total Extract Required Per Unit Brew

For the total extract (ΣE) required per unit brew in a constant OE process and given that beer OE = $OE_b = OE_w$ (in °P), then:
$$\Sigma E = V_{hl/brew} \times [0.9974 \times (P^{-1} - 0.00382)^{-1}] \tag{38}$$

For high-gravity brewing wort OE_w > beer OE_b; the original wort is brewed to a substantially higher OE_w than the prescribed for the final beer OE_b for the purpose of blending the beer with water to yield the designated beer. This practice has the following advantages:

- Reduced brewhouse volume demand
- Reduced cellar capacities in fermenting and lagering
- Improved product quality, if processed through the filtration as high-gravity product

High-gravity brewing processes require special facilities for supplying carbonated, oxygen-free, high purity dilution water.

Equation (38) equally applies to high-gravity (HG) brews for calculating the total extract required for the formulation.

B. Malt Requirement Per Unit Brew

For the malt requirement, kgs per unit brew:

a. All-malt brewing—one type of malt; no adjuncts used:

$$\text{Malt, kg per unit brew} = \frac{\Sigma E}{Y_m} \tag{39}$$

Where:

ΣE = Total extract per unit brew (kg)

Y_m = Yield of malt (%)

b. All-malt brewing, more than one type of malt; no adjuncts using malts a, b, c....n of percentage yield y_a, y_b, y_c....y_n; contributing extracts, as percentage of ΣE, the total extract, $E_a\%$, $E_b\%$, $E_c\%$ ... $E_n\%$.

For malt 'a', kg per unit brew is:

$$\frac{[\Sigma E \times E_a\%]}{y_a} \tag{40}$$

Kg per unit brew of other malts is calculated in the same manner using corresponding percentages and yields.

C. Kilogram per Unit Brew

The kg of each material in kg per unit brew and the formulations containing malt and adjuncts (a, b) are as follows:

$$\text{Malt kg per unit brew} = \frac{[\Sigma E \times E_m\%]}{Y_m} \tag{41a}$$

$$\text{Adjunct 'a' kg per unit brew} = \frac{[\Sigma E \times E_a\%]}{Y_a} \tag{41b}$$

$$\text{Adjunct 'b' kg per unit brew} = \frac{[\Sigma E \times E_b\%]}{Y_b} \tag{41c}$$

Where:

ΣE = Total extract per unit brew (kg)

Y_m = Yield of malt (%)

E_m = Extract contributing from malt as % of ΣE

Y_a = Yield of adjunct 'a' (%)

E_a = Extract contributing from adjunct 'a' as % of ΣE

Y_b = Yield of adjunct 'b' (%)

E_b = Extract contributing from adjunct 'b' as % of ΣE

The quantities of any additional materials are calculated in the same manner.

The ratios of extract-yielding materials are precisely stated in terms of the ratio of their individual total extracts. However, often the ratios are stated in terms of the gross weights of the materials. Gross weight ratios are to be avoided. This constitutes the calculation of the mash formulation.

D. Theoretical Yield

The theoretical yield of a formulation is:

$$\text{Yield} = \frac{\text{Total theoretical extract}}{\text{Total materials}} \times 100$$

$$\frac{\Sigma E}{\text{Total kg of materials}} \times 100 \tag{42}$$

E. Decoction Mash Calculations

Decoction mashing is usually employed to all-malt formulations. It can be applied in rare cases to a malt/adjunct formulation.

Decoction mashing withdraws increments of the total main malt mash; transfer such portions to a separate vessel, and heat this mash portion to a predetermined temperature, calculated so that when combined with main malt mash the resulting mash temperature will be precisely a specified value. This process may be repeated sequentially with the same mash as many as three to four times—withdraw, heat, combine, and repeat. In all the cases, the combined mashes are to be homogeneous, including thermal homogeneity. This simplifies the heat content calculations so that masses/volumes of the mash can be reduced to percentage extract or ratio. That is, the decoction portion may be mathematically describes as × percent of the total mash while the main mash is described as (100-x) percent. The equation (43) employs a percentage of the main mash. Alternately, exact mash volumes for each, main mash and decoction mash, may be used directly in the equation, replacing 'percent or %' with actual hl volumes.

To calculate the required decoction mash temperature to elevate the combined mash (main and decoction mash) to a specified temperature ($°C_{Mn+Dm}$):

$$°C_{Dm} = \frac{(100 \times °CMn+Dm) - °CMn \times (100 - X)}{X} \qquad (43)$$

Where:

$°C_{Dm}$	= Required temperature of the decoction mash
$°C_{Mn}$	= Temperature of the main mash
$°C_{Mn+Dm}$	= Temperature of the combined main and decoction mashes
X	= Decoction mash volume as percentage of the total mash

F. Wort Lautering Calculations

The lautering of wort from the mash consists of three operations:

1. Separation of solids of the mash from the liquid wort; establishing the mash as a filter medium (bed).

2. Collection of the first wort—the liquid component of the final converted mash—efficiently and completely.

3. Extraction of the residual first wort and soluble extract remaining in the grain 'bed' by sparging with water—termed 'sparging'.

Lautering is a critical process isolating the initial 'saleable' substrate. Efficiency is readily calculated and losses are unrecoverable.

The lautering process can be mathematically designed for optimum performance applying the modified Smith equations (Smith, 1943; Weissler, 1945).

Design of Total Sparge water Usage (Smith Equation S-1)

$$V_{SW} = V_{wkf} - V_{w-1} - V_{wsg} - V_e \tag{44}$$

Where:

V_{SW} = Total sparge water (hl) required

V_{wkf} = Wort (hl) at wort kettle fill

V_{w-1} = Total adjusted hl of first wort

V_{wsg} = Residual hl wort in spent grains

V_e = Wort kettle evaporation (hl)

Calculation of Adjusted First Wort Volume (Smith Equation S-2)

$$V_{w-1} = \frac{V_{w-1} \times E_{w-1}}{E_{SRW}} \times 100 \tag{45}$$

Where:

V_{w-1} = Observed hl of first wort delivered to wort kettle at $°P_{w-1}$

E_{w-1} = Equivalent kg extract per hl of V_{w-1}

ESRW   = Equivalent Kg of extract per hl of standard reference wort (designated kettle drop wort)

Residual Wort (hl) Remaining in Spent Grains (V_{wsg}) (Smith Equation S-3)

The fluid remaining in the mash (spent grain) at the end of lautering is expected to be of the same composition and concentration as that of the Glattwasser. Further, if the spent grain has been allowed to drain to a level of 'no free water' or very nearly so, the moisture is approximately eighty percent. Two equations are furnished:

1. Assuming eighty percent moisture in the spent grain as a constant

2. For variable moisture

<u>Note</u>: *In the process of brewing, after the wort has been drawn off from the mash and the grain has been sufficiently steeped therein, (i.e.) at the end of lautering, the said weak wort (low sugar-water) is allowed to run to waste. It is commonly called 'Glattwasser', or sugar-water.*

It is necessary to calculate the residual of each brewing materials remaining in the spent grain and then add the values to the total.

For one component, malt (m):

$$V_{wsg.m} = \frac{(W_m \times Y_m) - (W_m \times M_m)}{(100 - Msg) \times 257.97} \qquad (46)$$

Or where:

Msg = Constant = 80%

$$V_{wsg.m} = \frac{(W_m \times Y_m) - (W_m \times M_m)}{5159}$$

Calculate V_{wsg} for the other materials and add for the total V_{wsg}:

$$\Sigma V_{wsg} = V_{wsg.m} + V_{wsg.a} + V_{wsg.b} + \ldots\ldots + V_{wsg.n}$$

Evaporation During Kettle Wort Boiling (V_e) (Smith Equation S-4)

The volume of water evaporated during the boiling of wort in the wort kettle should be a designated, constant percentage of the volume of the kettle fill wort volume. The recommended volume of evaporation ranges from six to ten percent, with a brewery mean estimated to be eight percent (minimum recommended by traditional brewers):

$$V_e\,(\%) = \frac{(\text{hl wort kettle fill} - \text{hl wort kettle drop})}{\text{hl wort kettle fill}} \times 100$$

$$V_e\,(\%) = \frac{(V_{w\text{-}kf} - V_{w\text{-}kd})}{V_{w\text{-}kf}} \times 100 \qquad (47)$$

Extract Loss, Retained in Spent Grain as Glattwasser (Smith Equation S-5)

The residual wort remaining in the mash (spent grain) at the end of lautering is Glattwasser and is expected to be of the same composition and extract concentration as that of the last flowing wort to the wort kettle. The extract of this non-recovered Glattwasser represents a loss, approximated by the following equations:

$$E_{gw\text{-}loss} = V_{wsg} \times E_{gw} \qquad (48)$$

Where:

$E_{gw\text{-}loss}$ = Extract (kg) retained in spent grains

V_{wsg} = hl of Glattwasser retained in spent grains [see Eq. (46)]

E_{gw} = Extract, kg/hl in the Glattwasser of $°P_{gw}$

Lautering Rate—water phase transport through grain (mash) bed (Darcy's Law):

$$F = \frac{K\Delta PA}{uh} \tag{49}$$

Where:

F = Lautering rate, rate of water phase transport through the grain bed

K = Average bed permeability $(u^3 d_e^3) \times [(1\text{-}u)^2 \times 180]^{-1}$

d_e = Surface averaged particle size (effective diameter) = $\Sigma^{-1}(X_i/d_i)$

U = Bed porosity (wort volume/bed volume)

ΔP = Pressure drop through the mash bed

The valid generalizations related to lautering arc as follows:

1. The volume of the first wort should represent at least fifty percent of the total wort volume collected per brew.

2. The original extract of the first wort should be of a constant concentration (°P) throughout collection.

3. The original extract concentration (°P) of the first wort (high, medium, low) will reflect the type of mash employed (thick, medium, thin).

4. The Glattwasser should be ≤0.5 °P as collected at the end of grain bed drainage.

5. The spent-grain-retained Glattwasser volume should be approximately equivalent to the residual spent grain volume.

6. Plot the concentration of the wort (kg extract/hl) versus volume of wort to the wort kettle at regular intervals throughout the lautering.
   - First wort, three periods: beginning, middle and end of collection
   - Sparging, three equal intervals (of volume of sparge)—Glattwasser at the end of sparge and spent grains drained

7. Integrate the curve by function (f) or means of a polar planimeter. Graphical results supplement the Smith data and avoid repetitive calculations.

Calculation of Boiling Point of Water at Various Altitudes

A linear relationship exists between the boiling point of water and the geographical altitude of the test station. This relationship is given by the formula:

$$BP_w = [-1.07 \times A \times (304.8)^{-1} + 100.1] \,°C \tag{50}$$

Where:

BP_w	= Boiling point of water (°C)
A	= Altitude of test site (meters)

This equation is useful for cereal mash boiling, wort boiling operations, and heat balance calculations.

Ethanol (alcohol) Percentage By Weight

This procedure and calculation afford a reliable means of determining the ethanol content by weight of beer from the depression of the boiling point of water using simple and readily available inexpensive equipment.

The equipment consists of an ebulliometer, a differential thermometer graduated in 0.01°C, and Plato hydrometer. The boiling point of water is first determined and recorded. The boiling point of degassed beer and its real extract is then determined. The boiling point depression and the difference between the boiling point of the water and beer are calculated from the formula:

$$\Delta bp = bp_w - bp_b \tag{51}$$

Determine the corresponding apparent ethanol (E_a) value from Table 1 and add the real extract correction (C_{re}) of Table 2 as:

$$E = E_a + C_{re}$$

Where:

E	= Ethanol (alcohol) (% wt)
E_a	= Apparent ethanol (% wt) (from Table 1)
C_{re}	= Correction for real extract bp elevation (from Table 2)

Note: *The ethanol in solution lowers the boiling point of the solvent—water—while the real extract elevates the boiling point of water, and therefore elevates the correction coefficient.*

<u>*Boiling Point of Wort*</u> (°C$_{bp\text{-}wort}$):

$$°C_{bp\text{-}wort} = \frac{°C_{bp\text{-}water} + K_b (10^3 \times °P)}{(100 - P) \times MW_{we}} \tag{52}$$

Where:

$°C_{bp\text{-}wort}$ = Boiling point of the wort ($°C$)

$°C_{bp\text{-}water}$ = Boiling point of water ($°C$)

$°P$ = Plato of the wort

MW_{we} = Apparent molecular weight of wort extract (solute)

K_b = Molal boiling point constant (0.512)

The equation is necessary for precise heat balance calculations and comparison of process conditions at different altitudes.

Table 17.1: Relationship Between Boiling Point Difference and Percentage Alcohol By Weight

Difference between boiling points	Difference between Alcohol, % by wt	Difference between boiling points	Alcohol, % by wt	boiling points	Alcohol, % by wt
0.05	0.040	0.47	0.377	0.89	0.718
0.06	0.048	0.48	0.386	0.90	0.726
0.07	0.056	0.49	0.395	0.91	0.734
0.08	0.064	0.50	0.404	0.92	0.742
0.09	0.072	0.51	0.413	0.93	0.750
0.10	0.080	0.52	0.422	0.94	0.758
0.11	0.088	0.53	0.431	0.95	0.766
0.12	0.096	0.54	0.440	0.96	0.774
0.13	0.104	0.55	0.448	0.97	0.782
0.14	0.112	0.56	0.456	0.98	0.790
0.15	0.120	0.57	0.464	0.99	0.800
0.16	0.128	0.58	0.472	1.00	0.806
0.17	0.136	0.59	0.480	1.05	0.950
0.18	0.144	0.60	0.488	1.15	0.990
0.19	0.152	0.61	0.496	1.20	1.030
0.20	0.160	0.62	0.506	1.25	1.041
0.21	0.168	0.63	0.512	1.30	1.062
0.22	0.176	0.64	0.520	1.35	1.102
0.23	0.184	0.65	0.528	1.40	1.150
0.24	0.192	0.66	0.536	1.45	1.190
0.25	0.200	0.67	0.544	1.50	1.238
0.26	0.208	0.68	0.552	1.55	1.278

Table 17.1 (Continued)

Difference between boiling points	Difference between Alcohol, % by wt	Difference between boiling points	Alcohol, % by wt	boiling points	Alcohol, % by wt
0.27	0.216	0.69	0.560	1.60	1.326
0.28	0.224	0.70	0.568	1.65	1.366
0.29	0.232	0.71	0.576	1.70	1.414
0.30	0.240	0.72	0.584	1.75	1.454
0.31	0.248	0.73	0.592	1.80	1.502
0.32	0.256	0.74	0.600	1.85	1.542
0.33	0.264	0.75	0.608	1.90	1.582
0.34	0.272	0.76	0.615	1.95	1.622
0.35	0.280	0.77	0.623	2.00	1.670
0.36	0.288	0.78	0.631	2.05	1.710
0.37	0.296	0.79	0.638	2.10	1.758
0.38	0.304	0.80	0.645	2.15	1.806
0.39	0.312	0.81	0.654	2.20	1.854
0.40	0.320	0.82	0.662	2.25	1.902
0.41	0.328	0.83	0.670	2.30	1.950
0.42	0.336	0.84	0.678	2.35	1.998
0.43	0.344	0.85	0.686	2.40	1.046
0.44	0.352	0.86	0.694	2.45	1.092
0.45	0.360	0.87	0.702	2.50	2.130
0.46	0.369	0.88	0.710	2.55	2.180
2.60	2.23	4.70	4.40	6.85	7.00
2.65	2.27	4.75	4.45	6.90	7.07
2.70	2.31	4.80	4.51	6.95	7.14
2.75	2.36	4.85	4.57	7.00	7.20
2.80	2.41	4.90	4.62	7.05	7.27
2.85	2.46	4.95	4.68	7.10	7.34
2.90	2.51	5.00	4.74	7.15	7.41
2.95	2.56	5.05	4.79	7.20	7.48
3.00	2.61	5.10	4.85	7.25	7.55
3.05	2.66	5.15	4.90	7.30	7.62
3.10	2.72	5.20	4.96	7.35	7.68
3.15	2.76	5.25	5.02	7.40	7.75

Table 17.1 (Continued)

Difference between boiling points	Difference between Alcohol, % by wt	Difference between boiling points	Alcohol, % by wt	boiling points	Alcohol, % by wt
3.20	2.80	5.30	5.07	7.45	7.82
3.25	2.85	5.35	5.13	7.50	7.88
3.30	2.91	5.40	5.19	7.55	7.95
3.35	2.96	5.45	5.25	7.60	8.02
3.40	3.01	5.55	5.36	7.65	8.09
3.45	3.06	5.60	5.42	7.70	8.17
3.50	3.10	5.65	5.48	7.75	8.24
3.55	3.15	5.70	5.53		
3.60	3.20	5.75	5.59		
3.65	3.25	5.80	5.65		
3.70	3.31	5.85	5.71		
3.75	3.36	5.90	5.77		
3.80	3.42	5.95	5.83		
3.85	3.47	6.00	5.89		
3.90	3.52	6.05	5.95		
3.95	3.58	6.10	6.01		
4.00	3.63	6.15	6.07		
4.05	3.68	6.20	6.14		
4.10	3.73	6.25	6.21		
4.15	3.78	6.30	6.27		
4.20	3.84	6.35	6.34		
4.25	3.89	6.40	6.40		
4.30	3.95	6.45	6.47		
4.35	4.00	6.50	6.54		
4.40	4.06	6.55	6.60		
4.45	4.12	6.60	6.67		
4.50	4.17	6.65	6.73		
4.55	4.23	6.70	6.80		
4.60	4.29	6.75	6.87		
4.65	4.34	6.80	6.93		

Source: Juerst Ebulliometer Manual, Elimer and Amend, New York, 1993, p.3. Table1.

Table 17.2: Correct Factors for Real Extract

Real Extract or solids contained in beverage%	Extract factor to be added to alcohol% by wt as found in Table 1
3.0	0.043
3.5	0.050
4.0	0.058
4.5	0.065
5.0	0.072
5.5	0.079
6.0	0.086
6.5	0.094
7.0	0.101
7.5	0.108
8.0	0.115
8.5	0.122
9.0	0.129
9.5	0.136
10.0	0.143
10.5	0.150

Source: Juerst Ebulliometer Manual, Elimer and Amend, New York, 1993, p.6.

Hops α-Acid Usage and Yield

The quantity of hops used in a brewing formulation must include the:

1. Weight of the hops of each type (variety)
2. Specific α-acid content of each hope type
3. Rate of usage stated as total weight of α-acids per unit weight of wort extract; or weight of α-acids per unit volume of wort specific ˚P.

This procedure is necessary because the unit volume of the 'brew' cannot be preserved for reference, but the extract remains a reliable reference. The group of α acid is generally accepted as being the most important measurable component of the hops:

$$\text{Total } \alpha\text{-acid used (wt)} = (H_1 \times \alpha_1) + (H_2 \times \alpha_2) + \dots + (H_n \times \alpha_n) \qquad (53a)$$

$$\text{Hop usage rate (wt/wt)} = \frac{\Sigma\ \alpha\text{-acids}}{\Sigma\ \text{Wort extract}} \qquad (53b)$$

Conveniently expressed as α-acids (mg/kg, wort extract), or ppm (wt basis, mg/kg), or IBU ppm per beer of standard $°P = k$.

For material balance calculations, this system must be practiced throughout the process, worts and beers alike. In addition, the beer analyses require the actual concentration of the α-acids in mg/l—either value can be calculated when the other is known:

$$\text{Hop Yield (\%)} = \alpha\text{-acid yield (\%)}$$

$$= \frac{\Sigma\ \alpha\ \text{recovered}}{\Sigma\ \alpha\ \text{input}} \times 100 \qquad (53c)$$

For whole hops, the yield is usually low, in the range of thirty to fifty percent. Hop concentrates usually give much better yields.

Where:

α	= Hop alpha acids
$H_1, H_2, \dots, H_n$	= Weight of each variety in formulation
$\alpha_1, \alpha_2, \dots, \alpha_n$	= Respective percentage of α-acids per variety
$\Sigma\ \alpha$-acid	= Total α-acids by weight used (= input)
$\Sigma\ \alpha$ recovered	= Total α-acids by weight from wort analysis
Σ Wort extract	= Total extract, by weight, in wort to which hops are added
K_b	= A constant ($°P$), OE of reference

Kettle Wort, Evaporation Rate, and Percentage of Total Evaporation

$$\text{Evaporation Rate} = \frac{(V_{t1} - V_{t2})}{(t_2 - t_1)}$$

$$= \frac{\Delta V}{\Delta t} \qquad (54a)$$

Where:

V_{t1} = Volume in wort kettle at time t_1

V_{t2} = Volume in wort kettle at time t_2

$$\% \text{ Total Evaporation} = \frac{(V_1 - V_2)}{V_1} \times 100 \qquad (54b)$$

Where:

V_1 = Volume of wort at kettle 'fill'

V_2 = Volume of wort at kettle 'drop' (transfer)

The traditional percentage of total evaporation of kettle wort is eight to ten percent. Less than six percent of total evaporation is unacceptable.

Note: Fractional kettle volume measurements may not be convenient or precise in the case of a microbrewery, and need only determine the kettle fill volume (V_1) and it's OE_1 and the kettle wort OE_2 at the end of the wort boil. The kettle wort volume (V_2) at the end of boiling and the evaporated volume are calculated as:

$$V_2 = \frac{E_1 V_1}{E_2} \qquad (55)$$

Where:

V_2 = Calculated volume in wort kettle at the end of boil

E_1 = Extract (kg/hl) of wort at kettle fill

V_1 = Volume in wort when wort kettle is full

E_2 = Extract (kg/hl) of wort in kettle at the end of boil

E_1 and E_2 are calculated from their respective samples °Plato. The present evaporation volume is then calculated in the usual manner.

$$\text{Kettle wort: } \% \text{ Volume Evaporation} = \frac{(E_1 - E_2)}{E_1} \times 100 \qquad (56)$$

Removal of Wort, Volatiles During Wort Boiling

$$C_t = C_0 \times (1 - V_w)^{\alpha - 1} \qquad (57)$$

Where:

C_t = Concentration of volatiles after time 't'

C_0 = Concentration of volatiles initially

V_w = percentage evaporation of wort in time 't'

α = Relative 'volatility' of the volatiles as compared to water

Note: This equation was identified without reference as the 'Rawleigh Equation' in the original article. However, it is better recognized as an application of Raoult's Law where α is derived as follows:

$$p_1 = p_1 x_1 \qquad p_2 = p_2 x_2$$
$$p_1 = p_1 y_1 \qquad p_2 = p_2 y_2$$

Combining:

$$\alpha = \frac{x_1 (x_2\, y_2)}{y_1} = \frac{p_1}{p_2} \qquad (58)$$

Where:

α = Relative volatility of wort volatiles (x) to water (y)

x_1 = Mole fraction (x) in liquid

x_2 = Mole fraction (x) in vapor

x_1 = Mole fraction (y) in liquid

y_2 = Mole fraction (y) in vapor

p_1 = Total pressure for the system (p_2)

Average wort of ~12.5°P, vigorously boiled at 1 kg/cm² atmospheric pressure, with a total evaporation volume of ~ ten percent is found to have a volatiles reduction of about ninety percent. There are variations due to higher Plato concentration, high hop rates, and so on.

Basis of Hop Usage

Hops are added in brewing on the basis of iso- α acid concentration designed for the final beer. This value is usually expressed in terms of "IBU value, in ppm, mg of iso- α acids per liter of wort or beer". It is necessary to work this exercise carefully in reverse order starting with the precise IBU value desired in the final beer. Almost without exception, every formulation includes more than one variety of hops considering losses, and other factors that may alter the IBU concentration of IBU/OE ratio. Therefore, it is necessary to establish the percentage of α acid to be derived from each variety or the base the design on other special attributes such as aroma, but account for their potential α acid contribution.

The yield of iso- α acids (isomerized products) in wort kettle boiling is often termed as *hop utilization* signifying the yield of iso-α acids in percentage from the total α acids introduced.

Hop Utilization (HU) = % Yield

$$= \frac{\Sigma A_{iso}}{\Sigma A_{\alpha}} \times 100 \qquad (59)$$

Where:

ΣA_{iso} = Total iso-α acids in wort at kettle drop

ΣA_{α} = Total α acids added to the kettle wort and boiled

HU varies very slightly within a defined and controlled system. Generally processed hop products offer substantially higher yield (*for some products as high as ninety percent*) than leaf hops (*vary from thirty to fifty percent*).

IBU Standard Reference for Wort and Beer Profile

The unit 'IBU ppm' is useful and necessary, as is, for sensory descriptions, but it lacks the specificity required for material balance, losses, cost, and variation calculations in the process. A simple means of standard reference can be established by relating all IBU values throughout the process (IBU profile) to a standard, constant OE value (°P = k).

As an example: Wort is brewed to an original extract (SRW) of 15°P. The resultant beer is blended to two products:

1. At an OE of 12°P
2. At an OE of 9°P

The only means of relating these three products is through denominator of constant Plato or extract value. However, whichever is chosen is immaterial. Selecting the Plato of one of the products reduces the calculations to that of only two products instead of three. A preference is to select the original SRW values for both original extract and IBU. Illustrating the specifications and calculations in tabular form:

Material	°P	Kg extract/hl	f_{SRW}	IBU (SR)
Wort (SRW)	15	15.87	-	35
Beer 1	12	12.54	1.2656	(28)
Beer 2	9	9.30	1.7065	(21)

f_{SRW} is a factor relating the beer to the original wort and is derived by dividing the SRW wort kg extract/hl by the respective corresponding values of each beer:

$$f_{SRW} = \frac{E_{SRW}}{E_b} \tag{60}$$

$f\,(12 \rightarrow 15°P) = 15.87/12.54 = 1.2656$

Then:

    IBU (at $12 \rightarrow 15°P$) = 1.2656 IBU

Inversely:

    IBU (at $15 \rightarrow 12°P$) = 35/1.2656 = 27.7 IBU

This means that a wort of 15°P, containing 35 IBU when diluted with water without loss of any component, to 12°P, will have an IBU value of 27.7. The same relationship exists between the wort and its resultant beer blended to the same OE.

Whirlpool Wort Volume Flow

This equation is required for applications to the whirlpool calculations that follow.

$$V = V_e A_e \tag{61}$$

Where:

    V = Volume flow

    V_e = Mean entry velocity

    A_e = Entry cross sectional area

Because of the strong shear forces and eddy currents of the whirlpool wort. The entry velocity (V_e) should not exceed 5 m/sec.

Whirlpool Relative Difference Velocity (ΔV_r)

The difference between the entry velocity (V_e) and the circumferential velocity (V_c) should be minimal. This difference is identified as the relative difference velocity (ΔV_r) and is given as:

$$\Delta v_R = v_E - V_c \tag{62a}$$

Where:

    ΔV_r is minimal

Whirlpool Wort Angular Momentum Resulting from Wort Entry Jet on Cylindrical Volume

The simplified Denk equation is:

$$\llcorner M = pVV_eR \tag{62b}$$

Where:

  ⌐M = Wort angular momentum

  p = Wort density

  V = Wort volume flow

  V_e = Mean entry velocity

  R = Whirlpool radius

Limiting the entry velocity (V_e) obviously reduces the angular momentum (⌐M), which must be accompanied by increasing the wort volume flow (V) to attain effective particle sedimentation at optimum rates.

Whirlpool Wort Particle Sedimentation Rate

The following Stokes Law to "very nearly" approximate the settling velocity of particle in a centrifugal force field, such as a whirlpool, although particles never attain a "true terminal velocity".

$$u_1 = \frac{\omega^2 r \,(p_p - p)\, D_p^{\,2}}{18\mu} \tag{63}$$

Where:

  u_1 = Particle settling velocity (m/sec)

  ω = rate of rotation (rad (m)/sec)

  r = Radial distance

  p_p = Density of the particle (kg/m³)

  p = Density of the wort (kg/m³)

  D_p = Equivalent diameter of a solid spherical particle

  μ = Viscosity of the wort (kg/m/sec)

Wort Aeration Calculation

This equation is employed to determine the flow rate of air in the aeration of wort being transferred from the wort cooler to the fermenter:

$$q = \frac{(5.27 \times W \times O_2)}{(10^4 \times p)} \tag{64}$$

Where:

  q = Required air flow (SCFM)

  W = Wort flow (hl/h)

  p = Pressure of air being injected into wort (kg/m²)

  O_2 = Oxygen in ppm desired dissolved in the wort

In practice, it is desired that the wort transferred to the fermenter be saturated with oxygen. The saturated level is commonly 8–10 ppm. Application of this equation insures that saturation conditions are provided while avoided an excess of gas, which would cause undue foaming in the fermenter.

Floting Vessel Processing

The primary calculations of the flotation process are directed to the determination of the quantities of wort components removed, both desirable and undesirable constituents. Flotation practices are varied to the extent that generalized formulas are of little practical use. The parameters of common interest are:

1. The efficiency of 'cold break' removal
2. The efficiency of the separation of 'hot break' from the wort
3. Minimizing the time interval between the end of the kettle boil and the initiation of the flotation process
4. Wort drainage rate from the 'foam blanket'
5. Wort loss in the foam blanket
6. Loss of foam-forming and stabilization components
7. Hop iso-compound losses
8. Organisms lag phase and exponential growth phase characteristics

Yeast, Quantity Added (Pitched) to Wort for Fermentation
(Basis Number of Yeast Cells per Milliliter of Wort)

The most important aspect of inoculating wort with yeast is to introduce an inoculum that rapidly establishes bio-dominance. Therefore, the density of the initial inoculum is critical and should be designed for this purpose.

1. Based on the °P of the wort: A rapid, reliable, and easily applied rule-of-thumb calculation based simply on the degree Plato of the wort.
2. Based on wort-fermentable extract: A more precise and logical method in which the initial yeast cell count population is directly proportional to the mass of fermentable wort extract to be fermented in contrast to method (1) in which the yeast cell population is proportional to the total mass (fermentable and non-fermentable) wort extract.

Yeast, Initial Population in Wort on the Basis of °Plato

It has been empirically developed through experience and practice that successful and consistent fermentations preceded where the yeast population in the wort was in the ratio of approximately 1.25×10^6 yeast cells per ml per 1°P according to the equation:

$$Y_{c/ml} = \frac{(1.25 \times 10^6)}{°P} \tag{65}$$

Where:

$Y_{c/ml}$ = Number of yeast cells/ml of wort

$°P$ = Degree Plato of the wort

Yeast, Initial Population in Wort on the Basis of the Weight of Fermentable Extract of the Wort

$$Y_{c/ml} = (65 \times 10^4)(E_{w/hl} \, E_{\%f}) \tag{66}$$

Where:

$Y_{c/ml}$ = Number of yeast cells/ml of wort

$E_{w/hl}$ = Extract in kg/hl of wort

$E_{\%f}$ = Percentage fermentability of the wort extract

65   = A constant

Ethanol and Carbon Dioxide Production in Fermentation

1. The Gay-Lussac Equation: Originally presented by Gay-Lussac in 1815, it occurs in the Buchner cell free fermentation.

$$C_6H_{12}O_6 \rightarrow 2C_2H_5OH + 2CO_2 \uparrow$$

| 100 | 50 | 48.9 | grams |
| 1 | 2 | 2 | moles |

2. Yeast cell fermentation:

$$C_6H_{12}O_6 \rightarrow 2C_2H_5OH + 2CO_2 \uparrow + \text{yeast cell substract} + \text{by-products}$$

| 100 | 48.6 | 46.4 | ~ 3.0 | ~2.0 | grams |
| 0.1245 | 1.055 | 1.054 | | | moles |

The theoretical efficiency is approximately ninety-five percent.

Fermentation Efficiency, Approximate Production: Comparison of Wort and Beer Original Extracts

    This requires the original extract of the wort to be reliably documented for reference and a chemical analysis of the fermented beer.

$$Y_f(\%) = \frac{E_b}{E_w} \tag{67a}$$

Where:

E_w = Extract of wort (g/l)

E_b = Extract of fermented beer (g/l)

Y_f = Fermentation efficiency

Ethanol Yield, Basis

$$Y_f = \frac{A_b}{[(E_w - E_b) \times 100 \times 48.6]} \times 100 \tag{67b}$$

Where:

E_w	= Extract of original wort (g/l)
E_b	= Real extract of fermented beer (g/l)
Y_f	= Fermentation efficiency
A_b	= Ethanol (g/l) beer
48.6	= Constant, theoretical alcohol (g/100 g extract)

Fermentation Process Efficiency, Extract Reference

Using the wort and beer extract data of Eq. (67a) with the standard reference wort and standard reference beer volumes, proceed as follows:

$$Y_{fp}\,(\%) = \frac{V_{SRW} \times E_b\, V_b}{E_w} \times 100 \tag{68}$$

Where:

E_w	= Extract of original wort (g/l)
E_b	= Real extract of fermented beer (g/l)
V_{SRW}	= Volume of wort (hl) beer
V_b	= Volume of fermented beer (hl)
Y_{fp}	= Fermentation process efficiency [refer to Eq. (67a)]

Carbon Dioxide Produced in Fermentation

Volumes: The term *gas volume (s)* is used commonly in the beverage industries and means that the beverage or other liquid contains a given gas dissolved in or supersaturing the liquid in unit volumes of the given liquid.

For example, a liquid L is said to have a gas (G) content of n volumes. This means that any unit volume of the liquid the amount of gas in that unit volume is given by nV_L.

For a 650 ml bottle filled with beer carbonated to a level of 2.80 volumes of CO_2,

Volume of CO_2 = 650 × 2.8 = 1820 ml STP

The simplified equations suggest that glucose is the sole source of ethanol and carbon dioxide, when in reality all the fermentable hexoses, maltose, and maltotriose contribute to the pool of products.

In all-malt wort, these substrates are in a molar relationship as follows:

Worts converted with malt enzymes only:

Maltose >> Glucose >> Maltotriose >> Fructose >> Sucrose

Worts converted with additional α-amyloglucosidase:

Glucose >> Maltotriose >> Fructose (Maltose absent)

For all practical purposes, the fermentable extract of the light beer-type wort is almost glucose. The reaction mechanism proceeds solely through the hydrolysis of maltose and maltotriose to glucose and the direct fermentation of the total glucose. It must be recognized that this hydrolysis increases fermentable extract by approximately five percent of the maltose mass fraction.

Maltose	+	Water	→	2 (Glucose)	
$C_{12}H_{22}O_{11}$		H_2O		$2\ C_6H_{22}O_6$	
342		18		360	Kg
1		0.0526		1.0526	Kg
1		1		2	Moles

The theoretical yield of CO_2 from each of the two sugars is:

Glucose	+	Fermentation →	2 (Ethanol)	+	$2\ CO_2\uparrow$	
$C_6H_{22}O_6$			$2\ C_2H_5OH$		$2\ CO_2\uparrow$	
180			92		88	Kg
1			0.511		0.489	Kg

Maltose	+	Water →	2 (Glucose) + Fermentation →	4 (Ethanol) +	$4\ CO_2\uparrow$	
$C_{12}H_{22}O_{11}$		H_2O	$2\ C_6H_{22}O_6$	$4\ C_2H_5OH$	$4\ CO_2\uparrow$	
342		360		184	176	Kg
1		1.0526		0.538	0.515	Kg

Fermentation (enzyme fermentation, noncellular) values served as a reference for estimating loss to new cell mass in production. Yeast fermentation values are repeated:

1 kg Glucose → ~ 0.486 kg Ethanol ~ 0.464 kg CO_2 ~ 0.02 kg cell mass by-products

The theoretical equation is:

$$CO_2\ (kg/hl) = 0.189\ (E_w - E_b) \tag{69}$$

Where:

E_w = kg extract/hl in original wort

E_b = kg extract/hl in beer

Carbon Dioxide Recovered from Fermentation

There are many reasons for diligence in recovering carbon dioxide gas at highest practical level of efficiency. Trade literature suggests that as low as fifty percent recovery is acceptable, while responsible brewers have indicated as much as seventy percent recovery, processing low-purity gas. 'Recovery CO_2 gas' implies purity of 99.98 percent CO_2 and within control and management.

The volume of the gas contained in beer leaving the fermenter may be as high as 1.5 volumes and is definitely 'recovered'. It appears that the largest loss of CO_2 is from 'waste'—insufficient effort for optimum management. Free venting of fermenters and similar discharges are irresponsible practices.

Carbon Dioxide Content of Beer (Volume and Percentage by Weight)
Virtually all results of beer carbonation analyses in volume units because of the simplicity and directness of measurement and derivation of results. Often the accuracy of the results is sacrificed by neglecting the enhancement of corrections for 'headspace' and headspace 'air'.

For CO_2 percentage by weight using the conversion from volume:

$$CO_2, \% \ Wt. = \frac{(CO_2)_v}{(5.093 \times SG_b)} \tag{70}$$

For the volume of CO_2, it can be directly determined from the solubility charts (ASBC, 1993-94) or can be calculated from the value of CO_2 percentage by weight as:

Volume $CO_2 = 5.093 \times CO_2$, percentage weight $\times SG_b$ (71)

Beer Blending (Dilution)
To blend a beer with water to yield 100 hl of blended beer of a given ethanol concentration.

$$V_{hg} = \frac{(100 \times A_{hg})}{[(A_{hg}S_{hg}) - (A_{bb}S_{hg}) + A_{bb}]} \tag{72}$$

Where:

A_{bb} = Ethanol percentage of the blended beer (specification)

A_{hg} = Ethanol percentage of the high-gravity beer (by analysis)

S_{hg} = Specific gravity of the high-gravity beer (by analysis)

V_{hg} = hl of high-gravity beer required

Then, the hectoliter of water required is $(100 - V_{hg})$.

For example, a standard high-gravity beer of the following analysis:

Ethanol = 5.5 percent wt, specific gravity = 1.01175, blend this beer with water to yield 100 hl of beer with an ethanol = 3.50 percent wt. What is the required volume in hl of the high-gravity beer?

Using equation (72):

$$V_{hg} = \frac{(100 \times A_{hg})}{[(A_{hg}S_{hg}) - (A_{bb}S_{hg}) + A_{bb}]}$$

$$= \frac{(100 \times 3.50)}{[(5.50 \times 1.01175) - (3.5 \times 1.01175) + 3.50]}$$

$$= \frac{350}{5.5235}$$

$$= 63.37 \sim 63 \text{ hl of high-gravity beer}$$

Then volume of water required for blending is (100 - 63.37) = 36.63 ~ 37 hl

The blend ratio of high-gravity beer to water is:

$V_{hg} :: V_{H2O}$ in the example is 63 : : 37

The blend factor of high-gravity beer to water is:

$$f \, V_{hg} = 100 \text{ hl} \tag{73}$$

$$f = \frac{100}{V_{hg}}$$

In the equation:

$$f = \frac{100}{63}$$

f = 1.587

This signifies that one unit volume of the high-gravity beer properly blended with dilution water will yield 1.587-unit volume of the designed blended beer.

Losses of Extract in Brewery

In order to have a complete understanding of the process, and be able to control the technology and cost of operations, it is imperative that both accurate and theoretical material data be accurately known throughout the brewery operations. The simplest data are yield and efficiency of each processing stage and of the overall operation.

The material balance accounting procedure is necessary to insure quality, efficiency, and economy of operations as well as environmental and other regulatory conformance. This exercise should be undertaken for each product, process, and any alteration of formulation or process.

The material balance data must be carefully analyzed by personnel from quality control, process control, operations, and engineering. These groups should seek and expediently implement every opportunity to improve quality, yield, efficiency and economy.

The basic material balance parameter is the extract of the product throughout the process. This consists solely of the potential soluble extract of the brewing materials:

- Dissolved solids (solute) in the unfermented products
- Both the dissolved solids and ethanol in the fermented products

An unwritten guideline for brewery losses is that the total losses, from receiving materials to packaged beer removal at warehouse should be less than fifteen percent, and desirably less than twelve percent.

A further industry guideline is that the processing losses from brewing through packaging should be less than ten percent. It has been found and confirmed through the data of the breweries having capacity more than two million hectoliters, controlled losses to within 7.5 percent to ten percent throughout processing.

'Wort contraction' or 'wort shrinkage', which refers to the difference in hot and cold wort volumes, should not be taken as loss. Fluid volumes are temperature dependent and accordingly can be compared only at a specified temperature. Hot wort volume must be adjusted to a standard temperature for all calculations, preferably 20°C, corresponding to the temperature of the Plato. The difference in volume between 100°C and 20°C wort is approximately -4 percent, thus:

$$V_{20°C} = 0.96 \times V_{100°C}$$

Obscure or Hidden Losses

Obscure losses are those that are not readily evident and are difficult to identify. An example is the entrainment of ethanol in fermentation gas and ethanol component of all counter pressure headspace. This particular example also represents a volatiles loss. The loss of ethanol is a disproportionate component loss that is most important. Another common source of 'hidden' losses is the diffusion of beer into the water used to flush beer conduits. The water beer interface is never discrete and, unless there is a gross waste of beer, some quantity of beer contains added water while the discharge will contain some beer. This presents the dilemma of 'no apparent volume loss' and a hidden extract loss.

The precise determination of volumes, specific gravities, and pertinent compositional analysis of the reclaimed products is necessary for the material balance audit as well as in judging the value of the reclaimed product for reintroduction into the process. The extract of any quantity of a reclaimed product must be treated as a formulated extract and included in 'input' and 'output' calculations.

Wort and Beer Color—Estimation from Formulations

Wort and beer color is a direct function of the sum of the soluble coloring matter of the materials composing the product formulations.

The color of the beer in process is directly related to the original extract of the beer and proportional to the original extract of the wort. These relationships prevail provided there are no additives or non-standard alterations to the system, such as the use of activated carbon, addition of sulfites, or the introduction of color-contributing additives.

Color, specifications and analyses, should be a part of the all the materials of the formulation known to contribute color to the wort. The color value of each material should be stated in terms of SRM or EBC color units contributed by a given weight of water = ~n SRM units or 1 kg of material per 1 hl of water = ~m EBC units, or, for a Congress wort:

50 g malt per 450 g water = $(SRM)_{CW}$ color contribution

$EBC = SRM \times 1.97$

$SRM = EBC \times 0.508$

For most brewers, the metric relationship of the Congress wort value directly applied in the simplest and most convenient.

For example:

- 1 kg of pale malt yields an EBC of 0.33 in hl of wort,

- 1 kg of caramel malt yields an EBC of 6.58 in hl of wort,

- 1 kg of Munchener malt yields an EBC of 1.64 in hl of wort,

What is the color (EBC) of 100 hl of wort brewed from only?

1. Pale malt	: 1500 kg
2. Caramel malt	: 100 kg
3. Munchener malt	: 25 kg

Color value unit (CV):	= EBC of 1 kg in 1 hl
Color units:	= Color contribution of materials (kg × CV)
Color of wort:	= (Σ Color units) × (Volume of wort in hl)$^{-1}$

Then,

Σ Color units = (0.33 × 1500 + 6.58 × 100 + 1.64 × 25) = 1192.1

Malt	Color Value	× Kg	= Color Units
Pale	0.33	× 1500	= 493.1
Caramel	6.58	× 100	= 658
Munchener	1.64	× 25	= 41
Total			1192.1

The EBC (Color) of wort is $(1192.1) \times (100)^{-1} = 11.921 \sim 11.92$ EBC

Color of Product Resulting from Blending Beers

If two or more beers of different colors are blended, the resulting beer color is calculated by using the following equation:

$$EBC_p = \frac{[V_a(EBC)_a + V_b(EBC)_b \ldots + V_n(EBC)_n]}{[V_a + V_b \ldots + V_n]} \tag{74}$$

Where:

$(EBC)_p$ = EBC (color) of the blended product

$V_a, V_b \ldots, V_n$ = Volumes of the beers used in blending

$(EBC)_a, (EBC)_b \ldots, (EBC)_n$ = Respective color values

Color, High-gravity Beer Blending

In blending a high-gravity beer with water to a lower OE beer, the color, EBC, is inversely proportional to the dilution. The greater the dilution, the lower the color:

$$EBC_{bb} = \alpha \, (EBC)_{hg} \tag{75}$$

Where:

EBC_{bb} = Color of the final blended beer

EBC_{hg} = Color of the high-gravity beer

α = Reciprocal of the blend factor $(f_b)^{-1}$

For example: A high-gravity beer of 16°P (OE) and an EBC of twelve is to be blended with water to 10°P (OE) beer. What is the approximate color (EBC) of the blended beer?

Solution:

16°P = 16.9973 kg extract/hl

10°P = 10.3701 kg extract/hl

Blend factor = f_b = 16.9973 × (10.3701)$^{-1}$ = 1.6391

EBC of the blended product = 12 × (1.6391)$^{-1}$ = 7.321 = 7.3

Color of Beer, Color Intensity Increase by Cellar Additives, Required Amount of Additive

Sometimes, a brewer may wish to increase the color intensity of a cellar beer. The parameters in such cases are the color of the beer (EBC_b); the (V_b) of the beer to be treated (hl), the color value of the color additive $(CV)_{ad}$, EBC/kg/hl; W_{ad} is the weight of the additive required (kg); and the desired final color of the product, $(EBC)_p$.

$$W_{ad} = \frac{(\Delta EBC \times V_b)}{CV_{ad}} \tag{76}$$

Equation (76) is applicable for the use of malt extract with normal spectral absorbance in the range 415–425 nm range.

Buffering Capacity and the pH of Wort and Beer

The buffering capacity of worts and beers is important because of its effects on the variation or control of hydrogen ion, [H$^+$] when sources of hydrogen ions are increased or increased within the system.

The pH is important for the following reasons:

1. It affects the flavor of beer

2. As pH rises, the bitterness of beer increases

3. The susceptibility to biological instability is a direct function of pH

4. The flavor stability of beer is a direct function of pH

The Kirsop (1978) equation:

$$\text{Buffering capacity} = \frac{[H^+]}{\Delta [H^+]} \tag{77}$$

This means that the buffering capacity of wort or beer is equal to the number of added hydrogen ions (µg) divided by the resulting change in connection of the free hydrogen ions (µg). The Kirsop method and equation is recommended, though somewhat cumbersome to the uninitiated. The routine method is to titrate 1 liter of wort of decarbonated beer with 1N HCl, noting the pH after the addition of each small (0.2 ml) increment of the acid. The pH ranges of interest are: Wort 4.9–6.0; beer 3.9–5.0.

The 'buffer profile method' (Weissler, 1978) titrates the wort and beer in the same manner but notes the volume of 1N HCl required to yield definite incremental changes in pH (0.10 units). This method gives an insight into the nature of the buffering components of the worts and beers in addition to the gross buffer capacity, (Reference: 1N HCl equivalence is 1 ml = 1000 μg of H^+). The relationship between pH and $[H^+]$ is given in Table 3.

Table 3: Hydrogen Ion [H⁺] Equivalence to pH (Kirsop, 1978)

pH	[H⁺]	pH	[H⁺]	pH	[H⁺]
3.1	794	4.1	79.4	5.1	7.9
3.2	631	4.2	63.1	5.2	6.3
3.3	501	4.3	50.1	5.3	5.0
3.4	398	4.4	39.8	5.4	4.0
3.5	316	4.5	31.6	5.5	3.2
3.6	251	4.6	25.1	5.6	2.5
3.7	200	4.7	20.0	5.7	2.0
3.8	159	4.8	15.8	5.8	1.6
3.9	126	4.9	12.6	5.9	1.3
4.0	100	5.0	10.0	6.0	1.0

Buffer Capacity Index of Van Laer (1925)

$$\text{Buffer Index} = \frac{10}{°P\,(pH_1 - pH_2)} \tag{78}$$

Where:

pH₁ = pH of the original sample, wort or decarbonated beer
pH₂ = pH after the addition of 4 ml 1 N HCl to 100 ml of the same sample
°P = Original extract of the sample

In its simplest form, the use of SRW warrants omitting the Plato value and reading simply the difference between the two pH values. This then yields a relative buffer value.

Notes on buffering capacity:
1. The buffering capacities of worts and beers are similar
2. Factor affecting wort and beer pH and buffering capacity:
   a. Buffering capacity
   b. Proportion of malt used in the formulation
   c. Protein content of the malt
   d. Yeast vitality

e. Yeast excretion of acids (organic) and H^+

f. Use of oxygen in fermentation to increase attenuation (causes a decrease of beer pH)

g. Restricted yeast growth, low yeast pitching rate, low O_2 tension (cause a rise in beer pH)

h. To control or change the pH of beer by means of altering the wort requires a change of one-and-a-half to two times the change in the wort as that of the beer:

$$\Delta pH_{beer} = (1.5\text{-}2) \times \Delta pH_{wort}$$

i. Any enhancement of yeast activity, in general, is reflected in a lowering of the beer pH

j. Malt-soluble nitrogen increases the buffering capacity

k. There is good correlation between buffering capacity and the free and combined aspartic and glutamic acid contents of worts and beers

l. Brewing materials effects on buffering capacity are in decreasing order: malt >> wheat flour >> (barley, maize and rice) >> sugars

m. Concentrations of buffering agents are proportional to the OE of the wort/beer. OE must be considered in all buffer capacity calculations

Pasteurization

Pasteurization of beer remains a widespread practice in the brewing industry. Accordingly, any brewer having facilities for the pasteurization of beer has the resources for its management. The critical parameters in the process and computations are:

1. Temperature range of 49–60°C for the product, uniformly and optimally. Temperatures over 60°C are excessive.

2. The time in minutes that the product remains at each temperature increment in the range.

3. The biological population of the beer.

4. The slope of the thermal death time curve, the 'Z' factor (-10.83°C for beer).

5. The physiochemical properties of the beer.

6. The definition of the pasteurization unit (PU)—a temperature of 60°C for one minute.

7. The treatment of data, integration of the area enclosed by the time-temperature curve and its abscissa, using special lethal rate coordinate paper. The area expressed in $inch^2$, where one in^2 equals one pasteurization unit (PU).

Item (7) represents the computation of the degree of pasteurization and is expressed in degrees PU *equivalent* minutes at 60°C over the temperature range of 49°C to the maximum temperature of the beer during processing.

The same principle is applied to bulk pasteurization employing a high temperature short-time process (HT-ST), in which all conditions of flow rate, temperature, residence time, and heat transfer are constant and thus with all parameters controlled, the effect, PU, is predetermined and remains constant. The PU equivalent data for the elevated temperatures are available from equipment manufacturers.

Estimating the Effectiveness of a Pasteurization Process

Beers fermented with *Saccharomyces uvarum (carlsbergensis)* contain the enzymes invertase and melibiose derived from the yeast in measurable levels of activity. Both enzymes, characteristically, are inactivated by heat at a rate proportional to temperature and time. This inactivation occurs during pasteurization and the relative degrees of invertase/melibiose activities are an indication of:

1. Relatively high enzyme activity; non-pasteurized product
2. No enzyme activity; totally effective pasteurized product-possibly over-pasteurized
3. Intermediate level of enzyme activity; a relative degree of exposure to pasteurization ranging from borderline to effective

The evaluation of the process can be readily quantified, predicted and controlled by analysis and calculations by adding specific amounts in terms of activities of invertase and melibiose to standard beer; determining the activities, as prepared; subject multiple samples for variable time/temperature treatment; immediately chilled and assay samples of each level of treatment; and plot the data, thus establishing a relationship between the enzymes' inactivation temperature/time conditions. These data represent the reference of comparison when similarly prepared samples are exposed to the production pasteurization. But most important, the test results are no longer just a 'go-no-go' answer, but rather a good estimate of the degree of pasteurization.

The computation method follows:

1. Plot the experimental data, abscissa (X-axis) = time, ordinate (Y-axis) = 1n of enzyme activity, or for simplicity, a linear plot
2. Establish a 'specification' range of enzyme inactivation for each product (effective range, minimum and maximum levels, and so on)
3. Evaluate production performance on the basis of enzyme inactivation of production samples containing added enzyme to the same activity levels
4. Determine performance on the basis to the reference graph

Efficiencies:

Factory (Brewery) Efficiency

An output measure, measuring the utilization of paid shift hours compared to the rated capacity of the plant/line.

Adjusted Factory Efficiency

An output measure as above, when discounting shift hours lost due to external uncontrollable factors.

Operating Efficiency

An output measure, measuring the utilization of available productive hours compared to the rated capacity after discounting planned maintenance and cleaning (M&C) hours.

Machine Efficiency

An output measure measuring the machine availability—the availability is affected by unplanned downtime, speed and quality losses.

Planned Maintenance and Cleaning (M&C) Hours

A line on a 2x8 shift pattern is allowed a *maximum* of eight factory hours per week for planned M&C. A line on 3x8 is allowed twelve factory hours per week; on 4x8, this increases to sixteen hours. Overtime production hours *do not* increase this allowance.

The recorded M&C time will be the lesser of the actual time spent and the maximum allowed planned M&C time for a given shift pattern. For example, if the maximum allowed M&C time for a 3x8 shift pattern is twelve hours and the actual time used is ten hours then the recorded M&C time should be ten hours, while if a skeleton crew on over time is used to do the same M&C and it exceeds twelve hours, then the recorded M&C time should only be twelve hours (when using skeleton crews, the actual M&C time will normally exceed the allowed M&C time).

For multi-pack lines, the actual M&C time should be recorded on the largest pack size being filled in that production week.

Paid Factory Hours

The paid factory hours is the total plant utilization time spent for production on normal and overtime plus the actual time spent to do planned M&C to a maximum of the allowed planned M&C time (approximately ten percent of normal time) for a given shift pattern. Maintenance and cleaning using contracted labor or overtime by maintenance staff is also deemed as paid factory hours to a maximum of the allowed planned M&C time for a given shift pattern. Planned shutdown and mini shutdown times are excluded from the paid factory hours for efficiency calculations.

Paid Factory Hour Adjustments

Paid factory hour adjustments are factory hours lost due to uncontrollable external factors to the brewery. These adjustments must be negotiated and agreed with planning.

The adjustment criteria are common for all breweries:

- General shortage of empties (bottles, cans and so on)
- Planning-imposed idle time
- Industrial Relations stoppages
- Commercial shortage of CO_2
- Raw material problems, when it is a national issue
- General water and power failures, acts of God, and so on
- New projects and trials where the product is not for commercial use
- Planned shutdowns or mini shutdown outages
- Other legal and business requirements adjustments as agreed with planning:
  - Fire drills
  - Annual and monthly stock-takes

Service Stops

Service stops are times lost on a packaging line or a complete packaging hall due to factors external to the packaging hall but controllable by the brewery, namely:

- Filtration
- Beer out of spec
- Beer shortage
- Automation problems to supply beer
- Raw materials (quality and supply)
- Warehouse
- Bottle shortage (internal)
- Faulty pallets (missing crates)
- Forklift trucks
- Mixed containers
- Utility supply (internal brewery failures)
- Fire alarm testing and drills

Allowed Stops

Allowed stops are line downtime allowances as a result of specific operating requirements of a line, which could be brand, pack, shift pattern or technology related. Allowed stop time standards are reflected in the capacity standards for each line. The allowed stoppages are:

- Brand changes
- Pack changes
- Daily stock-takes
- Start-ups and shutdowns
- Cleaning/sterilization (due to the product incompatibility)

For multi-pack lines, the pack change over time should be recorded on the hours of the pack size being changed to.

In summary, the different hours are made up as follows:

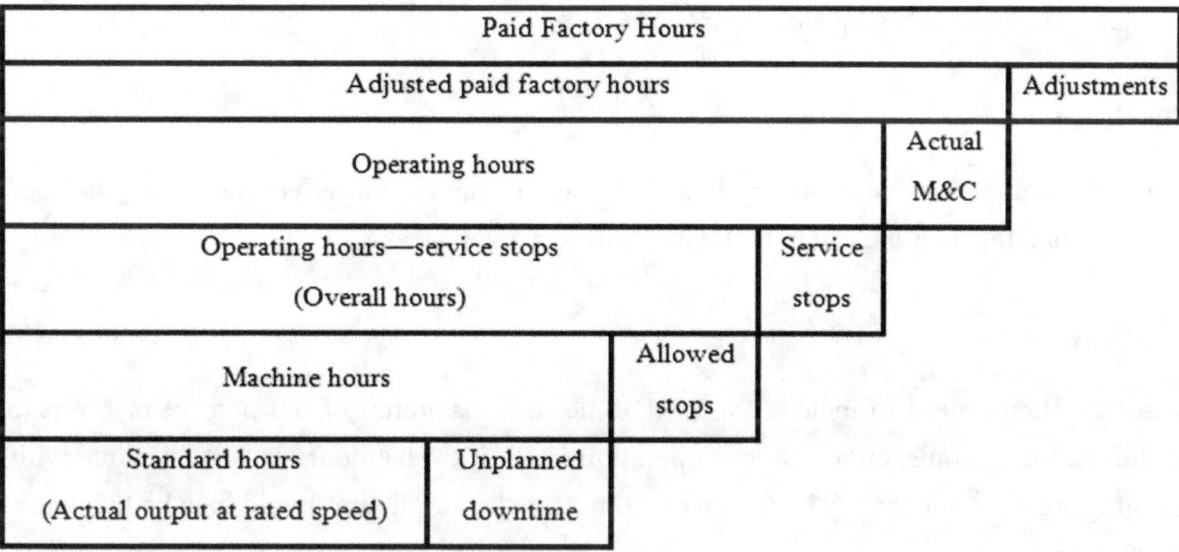

Paid Factory Hours				
Adjusted paid factory hours				Adjustments
Operating hours			Actual M&C	
Operating hours—service stops (Overall hours)		Service stops		
Machine hours	Allowed stops			
Standard hours (Actual output at rated speed)	Unplanned downtime			

Line Rating or Rated Speed

The line rating is the rated design speed of a packaging line as per the signed-off capacity standards. De-rating of a line might be required on a temporary basis when commissioning new equipment or products or when specific equipment or raw material conditions require a different line speed. De-rating of a line can only be allowed after obtaining a signed concession permit to run at a different rating. *All efficiency calculations will be based on the signed-off line ratings.*

Recording of Factory Hours and Operating Hours

'Factory hours' and 'operating hours' are used in the calculation of 'factory efficiency' and 'operating efficiency' respectively.

'Factory hours' refers to the shift hours paid for in a given period of time. Thus, for a week with 1x8 operation and no overtime, this would be forty hours; for a 2x8 operation, it would be eighty; for a 3x8 operation, it would be 120; for 4x8, it would be 168 hours. A 3x8 shift operation with twelve hours overtime on each Saturday and Sunday would be 144 factory hours. *No adjustment whatsoever is allowable.*

Operating hours makes an allowance for unproductive time on the line due to planned M&C. A *maximum* allowance, based on approximately ten percent of normal time factory hours, is permitted.

There are four basic situations how M&C can be executed:

1. M&C performed inside packaging shift hours
2. M&C performed by a packaging crew in overtime (at the weekend)
3. M&C performed by a skeleton crew outside packaging shift hours
4. M&C performed by a contract maintenance crew or a combination of contract labor and own labor outside packaging shift hours

Situation 1:

If the M&C is done inside the factory hours, the factory hours will reflect the labor time used for operating, maintenance and cleaning the line during a given week.

Situation 2:

The factory hours need to include the M&C time to a maximum of ten percent of the planned shift hours, for example, on a 120-hour operation (3x8), if twelve hours of M&C is done with the packaging crew outside factory hours (on overtime), the factory hours should then be 132 hours and *not* 120 hours.

Situations 3 and 4:

Where contract labor or skeleton crews or a combination is used for M&C outside factory production shift hours, then the factory hours will be the production shift hours plus the actual M&C hours worked to a maximum of ten percent of the production shift hours as allowed for according to shift patterns.

Compressed Weeks Paid Factory Hours:

When a packaging line are on a 2x8 and utilizing a compressed week shift pattern—(i.e.) 3x12+8—the planned M&C time must still be recorded within the total paid factory hours whether taken within the eighty hours or outside the eighty hours per week. For example, if the planned M&C is done in the *last* eight-hour shift, then the paid factory hours are eighty hours for that week. On the other hand, if the full eighty hours are used for production, and the M&C is done on the balance of that week's available time and say, the actual M&C time was five hours, then the paid factory hours are eighty-five hours.

Back-to-back Operation Paid Factory Hours:

When one crew is used to run two lines back-to-back—(i.e.) the first part of the week production is done on line A and the balance of the week is done on line B—then the paid factory hours by line must still reflect the actual time spent to do the weekly M&C for each line-up to the maximum allowed planned M&C time.

3x8 Sift Pattern Paid Factory Hours:

When a packaging line is on a 3x8 shift pattern, then the planned M&C time must still be recorded within the total paid factory hours whether taken within the 120 hours or outside the 120 hours per week. For example, if the planned M&C is done within the 120 hours for that week, then the paid factory hours are 120 hours for that week. On the other hand, if the full 120 hours are used for production and the M&C is done outside this time and say, the actual M&C time was eight hours, and then the paid factory hours are 128 hours. If the full 120 hours were used for production and the M&C was done on overtime with a skeleton crew and say, the actual M&C time was sixteen hours, then the paid factory hours are 132 hours as the lesser of the actual and the allowed planned M&C should be recorded.

Operating hours will be the factory hours minus the lesser of the allowed M&C hours and the actual M&C hours. As with factory hours, no other adjustments are made to this number. The allowed factory hours will be as per the M&C time as shown on the official packaging program agreed between the brewery and planning departments. The actual M&C hours taken will be communicated to planning the following day along with the other routine daily information.

Commissioning Time after Shutdowns

Commissioning time used for commercial production after a shutdown must be recorded as planned commissioning volume for that week. The commissioning time and volume must be negotiated and agreed with the planning department.

Pure Efficiency Calculations

These are the ratios between the actual volume produced and the volume that would have been produced in the quoted number of factory or operating hours had the line operated at its rated output.

Thus, if a line is rated at 200 hl/hr, and works on a 3x8 shift pattern, producing 18,000 hl in the week, its factory efficiency would be:

$$\text{Factory Efficiency} = \frac{18{,}000 \text{ hl}}{(200 \text{ hl/hr} \times 120 \text{ hrs})} \times 100 = 75\%$$

However, for operating efficiency, the operating hours will be determined by the amount of M&C time used. On a 3x8 shift pattern, twelve hours is the maximum allowance for M&C.

If twelve or more hours were used for M&C, the efficiency would be:

$$\text{Operating Efficiency} = \frac{18{,}000 \text{ hl}}{(200 \text{ hl/hr} \times 108 \text{ hrs})} \times 100 = 83.3\%$$

If only eight of the twelve hours were used for M&C, the efficiency would be:

$$\text{Operating Efficiency} = \frac{18{,}000 \text{ hl}}{(200 \text{ hl/hr} \times 112 \text{ hrs})} \times 100 = 80.3\%$$

Clearly, there is a degree of trust implied in using the lesser of the allowed M&C and the actual M&C as reporting the full allowance while actually using less will increase the resulting efficiency. Proper M&C planning and execution become more important and should drive the right behavior in this regard.

Again, it must be stressed that no adjustments to the hours whatsoever is allowed in this calculation.

Appendix

Glossary of Terms

Introduction

The definitions in this glossary are given as they would be applied within the brewing and malting industries. They should not be interpreted as a universal definition, such as the ones found in a standard dictionary.

The definitions have deliberately been explained in as simple a manner as possible, in order for any person working in the breweries, whether from a technical or non-technical discipline, to understand them easily.

A

Absorb

A process by which a material (absorbent) takes up another material, frequently from a solution, into the body of the material itself.

Acetaldehyde

An unpleasant by-product of fermentation that produces a 'green apple' or 'green avocado' flavor and aroma to beer.

Acid

Solids or liquids, with a pH less than 7, which imparts sourness or tartness (sharpness to the palate) to flavor.

Activated Carbon

Carbon from one of many different sources (such as coal or coconut shell) that absorbs flavors from water, carbon dioxide, or any other identified, compatible material.

Adjunct

A source of either fermentable or starchy extract that supplements the extract source for fermentation.

Adsorb

A process by which a material takes up another material, frequently from a solution, on to the surface of the material itself.

Aerobe

An organism that requires oxygen to survive.

Alcohol

Ethyl alcohol or ethanol—the main product of the brewer's yeast fermentation.

Aldehydes

A group of compounds found in beer, with similar chemical structures, which may give rise to stale flavors.

Ale

A beer made from top yeast and dark, flavored malts.

Alkalinity (of water)

The total alkaline constituents in water measured as calcium carbonate normally in ppm.

Alpha-Acid (*abbr.* α-acid)

Resinous constituent of hops from which a beer's natural bitterness is derived.

Amino Acid

Twenty similarly structured compounds found in beer, which are the result of total protein breakdown. They are important for yeast nutrition and collectively measured by FAN analysis.

Ammonia

The primary refrigerant in breweries, as it changes from gas to liquid form upon compression, and reverts to gas upon expansion.

Amylase

Enzymes that break down starch into sugars and other polymers that are smaller than starch.

Anaerobe

An organism that survives without oxygen. In some cases, oxygen acts as a toxin to an anaerobe.

Anion

A negatively charged component of a mineral dissolved in wort, beer or water.

Antioxidant

A compound added to beer (usually at filtration) that helps the product delay the staling process.

Apparent Extract (*abbr.* AE)

The gravity (in °P) of a beer at any particular stage of fermentation or in the final product.

Ash

The mineral content of the extract in beer or wort.

Assay

An analysis or series of analyses aimed at a desired laboratory determination.

Attenuation (Percentage)

Percentage of wort that is fermented at any stage. See Apparent (*abbr.* AA), Real (*abbr.* RA), and Limiting Attenuation (*abbr.* LA).

Autolysis

A process by which yeast cell walls rupture, thus spilling their protoplasm into the beer. It gives rise to 'yeasty' or even 'Marmite' type flavors.

B

Bacteria

Unicellular microorganisms of many different types, which do not have a nucleus. Can be rod- or spherical-shaped, and can be motile or non-motile.

Barley

The cereal from which malt is produced.

Beta-Gluconase (ß-glucanase)

Enzymes that break down ß-glucan (material found in cell walls of barley), which must work before protease and amylase can get at the protein and starch respectively.

Biomass

A term referring to the biological material dissolved in wort or beer.

Black Malt

Roasted malt used in the brewing of stout.

Bottom Yeast

Saccharomyces cerevisiae (alt. *uvarum*)—used in producing lager-type beers. It is called 'bottom', as the yeast settles to the bottom of the FV during fermentation.

Buffer

Materials dissolved in a liquid that minimizes pH change when acids or alkalis are added. FAN acts as a buffer during fermentation.

C

Calorific Value

A measurement of the food (energy) value of a beer given by the formula:

Kcal/100 g beer = 6.9(A) + 4(B-C)

Where,

> A = Alcohol, percent by weight.
> B = Real extract, percent by weight.
> C = Ash content, percent by weight.

Caramel

Product of intense amber color that is manufactured from cane sugar, and is used to darken beer and wort.

Carbohydrate

Range of compounds consisting only of carbon, hydrogen and oxygen that have food or structural value. All sugars, starches, dextrins and glucans are carbohydrates.

Casting

The process of pumping wort from the wort kettle to a whirlpool.

Caustic

A strong alkali with a sodium hydroxide or potassium hydroxide base. The former is used for cleaning vessels and mains in a brewery.

Centrifuge

A machine that separates solids from a liquid by a spinning action at high rpm.

Chill Back

The timeous chilling of an FV at the end of the fermentation process prior to racking to an SV.

Chill Haze

A reversible haze that appears when beer is chilled; the haze will disappear as the beer warms up.

Chloride

An anion of chlorine that adds sweetness and body to a beer, which is also associated with a condition known as stress corrosion of stainless steel.

Chlorine

A gas used as a sterilizing/disinfecting agent in water. Chlorine must be removed from water before using it as brewing liquor.

Coagulation

An irreversible condition of a protein due to structural changes within the molecule—the coagulated protein becomes insoluble and partly constitutes trub.

Cold Break

The portion of trub that is precipitated during wort cooling as it becomes super-saturated.

Coliform

A group of water-borne bacteria of mostly intestinal origin and distinguished by their ability to ferment lactose. They are frequently notorious wort spoilers.

Colloidal Stability

Measures the degree to which a beer is able to prevent haze formation over a period of time. The higher the colloidal stability of a beer, the longer its shelf life.

Conditioning

Beer in storage (maturation or lagering), or in ruh, may be said to be conditioning.

Consistency

Percentage of yeast cells in a liquid yeast slurry.

Conversion (of starch)

The breaking down of starch to fermentable and non-fermentable sugars and dextrins during mashing.

Conveyor

Any one of a number of kinds of moving machinery used to transport goods from one area to another.

For example:

Packaging conveyors for bottles, cans, cartons, and so on.

Material conveyors for malt, maize, sugar, and so on.

Copper

Old-fashioned name for a brewhouse kettle, so-called due to its copper construction.

Corn Starch

Pure, milled starch from a maize source that can be used as a brewing adjunct.

Counter Pressure

Pressure applied inside a packaging container (bottle or can) to allow filling to take place without loss of CO_2 or gushing.

Crown Cork

Generic name for bottle closure with a plastic/PVC lining.

Crowner

Machine for putting crowns on bottles.

Culture Yeast

The specific strain of yeast used by breweries for fermentation.

D

Decoction (brew)

Method of mashing such that a portion of the mash is transferred to a mash cooker, heated to boiling, and pumped back to the main mash vessel to effect required temperature changes. It was popular in continental Europe some years back, and helped produce acceptable worts from poorly modified malts.

Deflocculate

Process by which the aggregated components (yeast or protein) separate (for example, yeast deflocculation during acid conditioning).

Denaturation

An irreversible change in structure of a protein that renders it insoluble (for example, during hot break formation).

Dextrin

An intermediate polymer between non-fermentable sugars derived from starch.

Dextrose

Commercial name for glucose.

Diacetyl

Undesirable but natural off-flavor formed and removed during fermentation.

Diastase

An enzyme that produces maltose from starch and dextrins (alternative, ß-amylase).

Disinfection

The reduction in the number of microorganisms on a surface to a level that is acceptable in the context of the product being produced, the process being used and the standards to be achieved.

Doubling Up

A process by which a full, pitched FV is split into another FV and topped up with subsequent brews. It was used extensively by the Germans as a means for re-pitching yeast.

Drip Beer

Untreated, decanted beer from packaging destined for return to process.

Dusting (CO$_2$)

Light application of CO$_2$ into beer to help with air exclusion.

E

Efficiency

Broad term referring to productivity of an operation measured relative to expected or designed output.

Enzyme

A biological catalyst that can effect chemical changes in the malting and brewing process. Enzymes occur naturally in barley, malt and yeast. There are also commercially prepared enzymes available on the market, to assist the brewer.

Ester

Esters are distinctive flavor compounds produced as a by-product of fermentation. Common ones include ethyl acetate (solvent-like) and iso-amyl acetate (pear drop flavor).

Extract

The solid materials dissolved in wort or beer.

F

Fatty Acids

A group of naturally occurring chemicals found in malt, maize and yeast. Whilst some can be the precursor of staling compounds, trace amounts of fatty acids are necessary for yeast as a food source

Filler

Generic term for a machine that can fill either cans, bottles or both.

Filter

In breweries, the filter is usually responsible for the removal of yeast and other amorphous material from beer at the end of maturation (storage) en-route to the bright beer tanks.

Filtrate

The liquid phase emanating from a filter.

Finings

A brewing additive that aids in the sedimentation of amorphous matter and yeast, thereby improving filterability.

First Runnings (from a Lauter Tun)

The first wort to be run from a Lauter Tun before sparging.

Flocculation (of break or yeast)

The aggregating together of particulate matter in a suspension of wort or beer.

Flour

The finest fraction of crushed malt from a dry mill.

FOB

The foam that accumulates on wort or beer. Bottles ex-filler are 'fobbed' to exclude air from the package before crowning. CO_2 collection lines are fitted with fob traps to prevent beer carryover into the CO_2 system.

Fructose

A simple sugar derived from sucrose. Levels of fructose are occasionally not fermented in beer, leaving residual fructose (fructose block). Such beers are cloying and sweet.

G

Gas Volume (of beer)

The number of volumes of CO_2 gas that could theoretically be released from one volume of beer measured at 0°C and one atmosphere of pressure (NTP/STP).

Gelatinization

Process by which starch molecules are 'uncoiled' due to specific temperatures (for example, in a maize cooker), or during the malting process. Gelatinization is a prerequisite of liquefaction and solubilization of starch.

Generation

Term used to describe the number of times a yeast batch has been pitched.

Germicide

Generic name for a material that can kill microorganisms (for example, sterilizing agent).

Glucan

Alpha-glucan is starch, a macro-molecule of glucose units.

Beta-glucan is a structural carbohydrate found in the cell walls of barley endosperm. It is also a macro-molecule of glucose units, but structured differently from starch.

Glucose

A simple, fermentable sugar that is commercially known as dextrose.

Glycol

Short for propylene glycol—an anti-freeze added to water to depress its freezing point and used as a secondary refrigerant in breweries. Only food grade glycol may be used.

Gravity

The strength of extract in a wort or beer measured in °P.

Grist

The resultant mix of crushed malt after dry milling.

Grits

The resultant fraction of crushed malt after dry milling with particle size between husk and flour.
OR
The particle size of maize used as an adjunct.

Gushing

Condition of packaged beer resulting in massive release of CO_2 when the package is opened. This leads to the beer shooting out of the package.

Gypsum

Hydrated calcium sulfate—used as a brewing salt ($CaSO_4$ $2H_2O$).

H

Haemocytometer

Instrument that can be used in brewery microbiology laboratories for measuring yeast counts.

Hectoliter *abbr.* hl

Brewers unit equivalent to 100 liters or 0.1 m³.

High-gravity (brewing) *abbr.* HG

The brewing, fermenting, maturation and filtration of worts and beers at a higher gravity than the objective, and the subsequent dilution of the beer into BBT. This process achieves capacity requirements with the reduced purchase of expensive plant and infrastructure.

Higher Alcohols

Alcohols other than ethanol formed during fermentation.

Hopped Wort

Sweet wort after hops has been added in the kettle.

Hops

The plant *Humulus lupulus*, which bears a cone-shaped fruit containing glands that in turn contain bittering substances (resins) and flavoring substances (essential oils). It is the resin constituents which, when added to wort via the hops, that change chemically and impart beer's characteristic bitterness.

Hot Break

That portion of trub that is formed during kettle boiling and is separated in the whirlpool.

I

Infusion (mash)

A British system of brewing ales, whereby the mash conversion takes place at only one temperature (usually 63–65°C).

Inlet

The designed area where liquid enters a vessel.

Invert Sugar

An equal mixture of glucose and fructose sugars.

Invertase

An enzyme that splits sucrose into invert sugar.

Iodine Test

A test to ensure all starch has been converted in the mash tun. Samples of the mash are withdrawn and subjected to iodine solution. Dark blue/black coloration indicates the presence of starch.

Isinglass

A fining used to aid the sedimentation of yeast and other amorphous material in beer. It is normally used during maturation.

J

Jacket

A sealed shell around a vessel, through which a fluid can be pumped to control the temperature of the liquid inside the vessel. Steam or high-pressure hot water is normally used to heat the contents, whilst a glycol solution or ammonia is used for cooling.

Jetter

A device placed above the discharge of bottle filler that jets a powerful, but minute, stream of water into the bottle to cause it to foam. The foam thus excludes the air in the headspace of the bottle.

K

Keg

A vessel used to hold 'draught' (more correctly, kegged) beer. Most kegs are of 50-liter capacity.

Kettle

Brewhouse vessel where wort run-off from the Lauter Tun is boiled with hops. The boil accomplishes sterilizing, volatile stripping, haze-forming protein separation and color changes, amongst others.

Kettle Up

The beginning of the boil in the kettle.

Kieselguhr

Filter aid used to prevent a filter from blinding whilst producing bright beer.

Krausen

The German word for 'cauliflower', it describes the appearance of a firm white head on the top of the wort at the initial stages of fermentation.

'High krausen' describes the beginning of active, full fermentation, when yeast is in its most active state.

Kräusening

The addition of 'high krausen' to beer at the beginning of maturation at a rate of five to ten percent. This will allow for building up the CO_2 content during maturation by bunging under pressure.

L

Lactose

A non-fermentable, disaccharide sugar used for priming milk stout

Lag Phase

The time period between pitching of wort and the beginning of active fermentation.

Lager

A beer produced by bottom yeast. There are several types, including Munchner, Pilsener and Bock.

Lagering

Another word meaning 'maturation'. It originates from a German word that means 'resting'.

Last Runnings

The last wort of the lowest gravity that is washed out of a Lauter Tun using sparge liquor.

Lauter Tun

Brewhouse vessel (filter) with a slotted false floor used to separate sweet wort from its spent grains.

Limiting Extract (*abbr.* LE)

The remaining extract left behind in a fully fermented wort or beer.

Liquefaction

The solubilization of starch during the boil in a maize cooker.

Liquor

Brewer's term for product water.

Lye

Raw caustic soda.

M

Maize Grits

Maize kernels that have had their germ (embryo) removed, and have been milled to the particle size distribution required by the brewer for adjunct purposes.

Malt

The end product from malting barley. The barley has been modified in order to:
- Develop its enzyme system
- Break down endosperm cell wall structures
- Partially break down endosperm cell protein matrix

Maltings

Plant for malting barley.

Maltose

The most common fermentable sugar found in wort.

Maltose Syrup

A liquid adjunct consisting mainly of maltose.

Mash

The mixture of crushed malt, starchy adjunct and brewing liquor in a mash tun.

Mash Tun

Vessel used for extracting the required, desirable constituents from malt and starchy adjunct into the brewing liquor.

Methylene Blue

An indicator used to determine viability of yeast. Dead cells will stain blue, whilst viable (living) cells will remain unstained.

Microbiological Assay

A process of taking samples for microbiological contamination detection within a system to pinpoint the contaminating site.

Mill (for malt)

Dry mill:

Crushes malt through a series of two, four, or six rollers to a predetermined particle size distribution.

Conditioned:

Same as for dry mill, but the in-feed consists of a screw conveyor with hot water sprays to wet the husk. This allows a deeper bed in the Lauter Tun.

Steep conditioned:

- A wet mill, with normally two or four rollers, in which the malt is held dry in a bin and, during mashing-in, is subjected to a hot or warm water spray system in a chamber above the crushing rollers. The crushed malt is mixed into a mash beneath the rollers and pumped to the brewhouse.

Wet mill:

- Same as for a steep conditioned mill, except the malt is steeped in a wet bin above the rollers instead of passing dry into a steep chamber.

Mold

Uni- or multi-cellular filamentous and heterotrophic microorganisms that have a definite nucleus.

N

Nitrosamine

Naturally occurring carcinogenic (cancer-inducing) material found in malt as well as other foods such as bacon and cheese. Thus, its content is strictly controlled by indirect firing of the kiln during malting.

Non-fermentable Sugars

Any sugars in a wort or beer that cannot be fermented by culture yeast under ANY circumstances. They will always be left as part of the real extract of beer. For example, malto-tetraose (naturally occurring), and lactose (added as priming to milk stout).

Normal Gravity (*abbr.* NG)

A wort that is brewed at the required gravity in-package, that is, it is not diluted as in high-gravity

Nose (of a beer)

The smell of the beer.

O

Off-flavor

A naturally occurring flavor in beer at a stronger than desired level (for example, DMS, diacetyl, and so on).

Original Gravity (*abbr.* OG)

The strength of wort as it is collected, normally measured in °P. It can be calculated back from beer by measuring its apparent extract and refractive index.

Oxidation (of beer)

Process by which a beer develops stale flavors or precursors to stale flavors, either during processing or in-package. High oxygen (air) contents and warm temperatures accelerate this process.

P

Palate

The taste of a beer.

Paraflow

A plate heat exchanger; the one responsible for cooling wort en-route to fermenter.

Pasteurization

A process of heating beer to specified temperatures in order to render microorganisms incapable of further growth or activity. The degree of pasteurization is a time/temperature relationship, both factors being directly proportional to it.

Pasteurization Unit (*abbr.* PU)

An arbitrary scale defining the lethal 'rate' from holding beer at 60°C for one minute. Lethal rates at other temperatures vary exponentially with the temperature at 60°C.

For example, lethal rate for one minute at:

55°C	=	0.19 P.U.
59°C	=	0.72 P.U.
61°C	=	1.40 P.U.
64°C	=	3.70 P.U.

Pasteurization is a necessary evil, as it also accelerates the staling of beer. The importance of not over-exceeding 60°C thus becomes evident when looking at the exponential function.

Pathogen

Any microorganism that may lead to illness in humans.

Peptide

A broken down product of protein consisting of a chain of amino acids.

Peroxide

A compound that is both an extremely strong oxidizing agent and sterilizing agent.

Phenol

A compound that can taint beer severely (and has done so), despite its relatively minute concentrations. The two main sources are water (TCP-type taint), and a particular wild yeast (an oil-of-cloves-type taint).

Phosphoric Acid

The acid used for pH adjustment in the brewhouse. It can also be used as the base of an acid detergent used for cleaning and de-scaling vessels.

Pitching

The operation of adding yeast to wort during collection.

Polyclar at (*abbr.* PVPP)

A commercial name for polyvinyl polypyrollidine—a process aid used to enhance the haze stability of beer.

Polymer

A compound of very high molecular weight consisting of many repeating units of the same, or different species of molecules. For example, starch is a polymer of glucose.

Powdery

A term given to yeast that only flocculates slowly—that is, the yeast remains in suspension.

Precursor

A molecule that undergoes chemical changes to give rise to another known compound. For example, alpha-aceto lactate is the precursor of diacetyl.

Pre-Masher

Apparatus designed for mixing dry malt grist with water at mashing-in.

Primings (sugar)

A sugar solution added to beer at filtration, introduced to sweeten the beer.

A non-sweet, non-fermentable sugar (lactose) is added to the kettle for milk stout to increase the body of the beer. The lactose is also known as primings.

Propagation

The process by which a new culture yeast is grown, initially in the laboratory, and finally in a dedicated plant.

Protease

Enzymes that break up proteins into peptides and amino acids.

Protein

A macro-molecular compound made up of various amino acids. There are many different types, some of which are responsible for haze formation in beer, and others that are largely responsible for the natural foam.

Q

Qualitative Analysis

Any one of a number of laboratory analyses designed to show the PRESENCE of a constituent in raw material, wort or beer.

Quantitative Analysis

Any one of a number of laboratory analyses designed to measure the AMOUNT of a constituent in raw material, wort or beer (for example, air content).

R

Racking

The process of moving beer from one vessel to another, for example, FV to SV and BBT to draught beer keg.

Ranking

A procedure by which a series of samples are listed from best to worst, or *vice versa*.

Real Extract (Percentage) (*abbr.* RE)

The gravity (in °P) of dissolved solids in a beer at any stage of fermentation discounting the gravity effect of volatiles, especially alcohol.

Roasted Malt

Malt that is heated to high temperature and then cooled in cold water. This produces marked color and flavor formation. Roast malt is used only in stout.

Rollers

A pair of crushing cylinders in a mill.

Ropy

A serious microbiological growth in beer that produces long, visible filaments.

Rouse

The blowing of gas through or the mechanical agitation of a fluid, in order to mix the fluid contents thoroughly.

S

Saccharometer

Special hydrometer used to measure the strengths of worts or sugar solutions; can also be used to determine the apparent extract of beers in process and in-package.

Saccharomyces Carlsbergensis

An old scientific name of lager culture yeast.

Saccharomyces Cerevisiae

An old scientific name for an ale culture yeast. This has been reclassified to include all brewing yeasts.

Saccharomyces Uvarum

An old scientific name of lager culture yeast.

Salts (brewing)

Additives used during mashing and wort boiling—used to adjust the calcium, chloride or sulfate content of the medium.

Most usable salts are:

- Calcium sulfate (gypsum)
- Calcium chloride
- Potassium sulfate
- Potassium chloride

Sample Cock

Device with a valve used for taking a sample of wort or beer from a vessel or main, usually for analytical or tasting purposes.

Satiating

Flavor term describing a beer as being too filling or cloying.

Secondary Fermentation

The very slow fermentation that occurs in beer after the main fermentation, during ruh or maturation, or after Kräusening.

Sedimentation

The settling of flocs from suspension (for example, flocculated yeast sediments to the cone of a fermenter).

Sight Glass

Either a transparent gauge on a vessel, or a glass insert in a wort/beer/yeast main.

Silica Gel

An additive that adsorbs haze-forming proteins, thus enhancing the colloidal stability of a beer.

Silo

A large storage bin used for storing barley, malt, maize, spent grains, and so on.

Slurry

The stirred or mechanically agitated insoluble additive held in suspension in a dosing vessel.

Sparge Liquor

Brewing liquor used for washing out the available extract from spent grains in the Lauter Tun once the wort (first runnings) has been run-off.

Sparging

The process of washing out extract from spent grains in the Lauter Tun using sparge liquor.

Specific Gravity

The ratio of the mass of a substance of defined volume compared with the mass of an equal volume of water.

Specification

An allowed quantitative range around a parameter with a target value.

Spent Grains

The husks, acrospires and other materials left in grist after the extract has been washed out in the Lauter Tun. It is an important by-product.

Spray Ball

A hollow ball fitted at the end of a CIP line in a vessel, perforated specifically for the design of that vessel, and used to distribute detergents and sterilant in order to obtain a sanitary condition.

Stabilizing Agents

Any additives added to beer in order to improve its colloidal or flavor stability.

Starch

The major carbohydrate food source in malt and maize that is ultimately utilized by the yeast to produce alcohol and carbon dioxide during fermentation.

Sterilization

The complete elimination of all living organisms from a surface or vessel.

Storage

Specifically 'of beer in process'—a term used to describe the maturation or lagering stage. A storage vessel (SV) is a maturation or lagering vessel.

Sucrose

The chemical name of ordinary white cane sugar.

Sugar

A term used to describe a host of carbohydrate compounds of varying complexity, but generally not of macro-molecular size. Some are fermentable, whilst others are not.

Sulfate

A salt of sulfuric acid which, when added to brewing liquor, enhances astringent dry flavors in the resultant beers.

Sweet Wort

Wort before hops is added at the kettle. For example, the wort run-off from the Lauter Tun.

T

Taint

An unnatural flavor induced into beer from an external source (for example, raw material, sundry material, or plant). The flavor might be either present at its source, or the result of chemical activity with a beer or raw material constituent (for example, phenolic taint).

Tannic Acid

A specific additive that combines with certain proteins, normally during maturation, and thus enhances colloidal stability. That is, minimizes later haze formation in the packaged product

Tart

A pleasant, acidic flavor found in beer when it is at the required level. The flavor is similar to that of cream of tartar.

Ton

A metric ton, equivalent to 1,000 kg.

Top Yeast

Saccharomyces cerevisiae—used in producing ale-type beers. It is called 'top' as the yeast rises to the surface of the FV during fermentation.

Topping Up

A term used to describe an incremental stage of wort addition within one vessel during yeast propagation procedures.

Treatment

A generic term describing the effect of adding colloidal stabilizing materials to beer in process.

Trub

Amorphous material made up mainly from proteins and protein-tannin complexes that precipitate and flocculate during the brewing process.

U

Underback

An intermediate holding vessel used for sweet wort storage whilst waiting for the kettle to become available during the brewing purpose. Also known as the 'pre-run' vessel.

V

Viability

The percentage of living yeast cells or barleycorns in a population (for example, for yeast: propagation plant, pitching yeast, and so on).

Vitality

A loose term describing the health or fitness of a population of barleycorns or yeast cells.

Volatiles

Wort: A range of naturally occurring constituents that must be effectively removed from the wort during kettle boiling.

Beer: A totally different range of naturally occurring constituents, produced during fermentation, the levels of which must be controlled in order to produce the required flavors as per brand requirements.

W

Wild Yeast

Any yeast present in product or 'product-in-progress', other than culture yeast.

Wort

The solution of extract in brewing liquor that is fermented by yeast to produce beer.

Y

Yeast

Microorganisms containing a nucleus; there are many different types.

Yeast Count

The number of yeast cells per unit volume of beer or wort, usually expressed as the number of millions of cells per ml.

References

1. *Malting & Brewing Science* (2nd Edition), Vol-1, Briggs, Hough, Stevens, Young.

2. *The Practical Brewer* (2nd Edition), Harold M. Broderick

3. *A History of Beer and Brewing*, Ian S. Hornsey

4. *Barley: Production, Improvement, And Uses*, Steven E. Ullrich

5. *Brewing Microbiology* (2nd Edition), F. G. Priest & I. Campbell

6. *Brewing Yeast Fermentation Performance* (2nd Edition), Edited by Katherine Smart

7. *Handbook of Brewing*, Edited by William A. Hardwick

8. *Handbook of Brewing*, Edited by Fergus G. Priest & Graham G. Stewart

9. *Handbook of Brewing*, Edited by Hans Michael Eblinger

10. *Briggs of Burton*—Feature Articles

11. *Brewing Science & Practice*, Dennis E. Briggs, Chris A. Boulton, Peter A. Brookes and Roger Stevens

12. *The Brewer International*, Technical Summary (42), April 2002

13. *Brewing And Beverage Industry International*, No: 5/2011

14. *Brewer's Guardian*, Volume 140, No: 5 September/October 2011

15. *Perry's Chemical Engineers' Handbook* (7th Edition), Robert H. Perry & Don W. Green

16. *Handbook of Waste and Wastewater Treatment Plant Operations* (3rd Edition), Frank R. Spellman

17. 'Strict Hygiene Regime is Essential for Trouble-Free Bottle Filling', *Brewers' Guardian* 125 (8), O'Rourke, Tim. 1996

18. 'The Importance of Filtration in Producing Quality Beers', *Brewers' Guardian* 120 (1), Bennett, Keith. 1991

19. 'Practical Wastewater Pre-treatment Strategies for Small Breweries', *MBAA Technical Quarterly* 39(1), Ockert, K. 2002

20. 'Wastewater Minization and Effluent Disposal of a Brewery', *MBAA Technical Quarterly* 30, Watson, C. 1993